Prais

"What C-Level executives read
decisions. Timeless classics for
Manager-Corporate Marketing Communication, General Electric

"Want to know what the real leaders are thinking about now? It's in here." - Carl Ledbetter, SVP and CTO, Novell, Inc.

"Priceless wisdom from experts at applying technology in support of business objectives." - Frank Campagnoni, CTO, GE Global Exchange Services

"Unique insights into the way the experts think and the lessons they've learned from experience." - MT Rainey, Co-CEO, Young & Rubicam/Rainey Kelly Campbell Roalfe

"A must-read for anyone in the industry." - Dr. Chuck Lucier, Chief Growth Officer, Booz-Allen & Hamilton

"Unlike any other business books, *Inside the Minds* captures the essence, the deep-down thinking processes, of people who make things happen." - Martin Cooper, CEO, Arraycomm

"A must-read for those who manage at the intersection of business and technology." - Frank Roney, General Manager, IBM

"A great way to see across the changing marketing landscape at a time of significant innovation." - David Kenny, Chairman & CEO, Digitas

"An incredible resource of information to help you develop outside the box..." - Rich Jernstedt, CEO, Golin/Harris International

"A snapshot of everything you need to know..." - Larry Weber, Founder, Weber Shandwick

"Great information for both novices and experts." - Patrick Ennis, Partner, ARCH Venture Partners

"The only useful way to get so many good minds speaking on a complex topic." - Scott Bradner, Senior Technical Consultant, Harvard University

"Must-have information for business executives." - Alex Wilmerding, Principal, Boston Capital Ventures

Inside the Minds

The critically acclaimed *Inside the Minds* series provides readers of all levels with proven business intelligence from C-level executives (CEO, CFO, CTO, CMO, partner) from the world's most respected companies. Each chapter is comparable to a white paper or essay and is a future-oriented look at where an industry/profession/topic is heading and the most important issues for future success. Each author has been carefully chosen through an exhaustive selection process by the *Inside the Minds* editorial board to write a chapter for this book. *Inside the Minds* was conceived in order to give readers actual insights into the leading minds of business executives worldwide. Because so few books or other publications are actually written by executives in industry, *Inside the Minds* presents an unprecedented look at various industries and professions never before available.

INSIDE THE MINDS

Biotechnology Deals

Key Strategies for M&A, Financings, and Partnerships in the Biotech Industry

ASPATORE BOOKS

BOOK IDEA SUBMISSIONS

If you are a C-Level executive or senior lawyer interested in submitting a book idea or manuscript to the Aspatore editorial board, please email authors@aspatore.com. Aspatore is especially looking for highly specific book ideas that would have a direct financial impact on behalf of a reader. Completed books can range from 20 to 2,000 pages – the topic and "need to read" aspect of the material are most important, not the length. Include your book idea, biography, and any additional pertinent information.

SPEAKER SUBMISSIONS FOR CONFERENCES

If you are interested at giving a speech for an upcoming ReedLogic conference (a partner of Aspatore Books), please email the ReedLogic Speaker Board at speakers@reedlogic.com. If selected, speeches are given over the phone and recorded (no travel necessary). Due to the busy schedules and travel implications for executives, ReedLogic produces each conference on CD-ROM, then distributes the conference to bookstores and executives who register for the conference. The finished CD-ROM includes the speaker picture with the audio of the speech playing in the background, similar to a radio address played on television.

INTERACTIVE SOFTWARE SUBMISSIONS

If you have an idea for an interactive business or software legal program, please email software@reedlogic.com. ReedLogic is especially looking for Excel spreadsheet models and PowerPoint presentations that help business professionals and lawyers achieve specific tasks. If idea or program is accepted, product is distributed to bookstores nationwide.

Published by Aspatore, Inc.
For corrections, company/title updates, comments or any other inquiries please email store@aspatore.com.

First Printing, 2005
10 9 8 7 6 5 4 3 2 1

ISBN 1-59622-133-X

Library of Congress Control Number: 2005930118

Inside the Minds Managing Editor, Leah M. Jones, Edited by Eddie Fournier, Proofread by Brian Denitzio

Biotechnology Deals

CONTENTS

Successful Strategies for Biotech Deals

Lee R. Brettman, M.D., FACP
Chief Executive Officer and Director
Dynogen Pharmaceuticals, Inc.

Successful Strategies for Biotech Deals

Biotech deals differ from other types of deals in that the payouts generally take longer. Developing a biotech product can take up to ten years, whereas product development in the software industry is much shorter.

For a biotech deal to be successful, you need to be clear about what types of deals you want to be involved in, as opposed to just focusing on how badly you want to get a deal done. You want to look for a deal that fits into your company's strategic plan and that will help build your company's value. A startup company should look at deals where it can best leverage its background and experience, preferably in markets where an unmet need exists and there is limited competition.

You also want to look at whether there will be baggage you will have to deal with down the road. You want to make sure that in your eagerness to clinch a deal, you don't overlook a price that will have to be paid later. One of the most common mistakes in in-licensing a compound is focusing on near-term components of the deal, such as upfront payments and early milestones, and not paying sufficient attention to royalty structure around sales. For instance, if you will need a strong partner for the large Phase III approval trials and for marketing the drug, a heavy royalty burden to the original licensor plus other royalties that might be necessary for freedom to operate, such as access to formulation technology, can make it very difficult to get a meaningful return on your investment.

For example, you do a deal to in-license an early-stage compound from another company in order to build your product development portfolio. As part of getting the licensing deal done, you agree to pay the licensor a double-digit royalty. In the short term, you have an attractive compound in development, but in the long term, the stacked royalties you will have to pay the licensor, plus formulation or other technology royalties, may make it difficult, if not impossible, to recoup your investment in development, let alone make a meaningful return on that investment. When you have to cut the pie again to share the value with a marketing licensee, the deal may not be so appealing.

Usually in the most successful deals, both sides get a meaningful return on their investment. In the biotech and biopharmaceutical industry, collaboration between licensors and licensees is required to successfully develop a product. There must be sufficient incentives built into the deal so your partner is motivated to continue making contributions. That can be a very tough balance to strike.

We try to take that approach at our company. We have extensive expertise in neurobiology. This area of biology deals with the nervous system and, in the case of our company, how the nervous system controls bladder and bowel function. Because of our focus on neurobiology, we have insight into which biological targets and pathways are most important for modulating bladder and bowel function. We turn to our company pharmacology experts so we can conduct studies validating these biological targets. After identifying the targets we are interested in, we search various databases that can provide information regarding compounds that have the appropriate receptor binding profile — in other words, they bind to receptors we want to turn on or off, thereby producing the desired activity. Typically, we synthesize the identified compound and study it in our pharmacology models. Based on the results, we then select the compound we are interested in, build a patent estate around it based upon our neurobiology expertise, and contact the company that owns it. If no one owns the compound because it is in the public domain, we must identify novel ways of using that compound to protect its use in order to go after it.

Calculating the Return on an Investment

When looking to secure the rights to a compound, you need to do the following: conduct a worldwide search and find the company that owns the compound, assess the market opportunity, look at the potential market impact of the product, determine potential peak sales, and determine the cost of developing and launching the product. You have to look at the likelihood of success, as well as technical hurdles that will have to be overcome in developing a product. All of these things contribute to the value of a deal.

In terms of licensing compounds, you usually don't know how much room there will be for negotiations until you have had a few meetings. Your first step is to identify who will be part of your negotiating team and what their roles will be. You also need to determine up front what you think constitutes a reasonable deal and what you consider to be deal-breakers. These deal-breakers can be onerous financial terms or unreasonable requirements for development diligence.

As mentioned earlier, royalties a company would have to pay to a licensor are a critically important financial issue. There is a limit to the level of royalties you can pay for an early-stage program. The licensee will need to make a substantial investment in developing the drug and obtaining marketing approval. If you have to pay high sales royalties, you are not going to get a return on your investment and build value in your company.

Risk Management

Developing drugs is an extremely risky endeavor. Only one out of eleven drugs that go into clinical trials ever reaches the marketplace. We try to evaluate and reduce risk in several ways:

(1) We apply our understanding of neurobiology and our pharmacology expertise to identify rational uses of existing compounds in our areas of interest (i.e., in GU and GI diseases, such as overactive bladder and irritable bowel syndrome). These disorders affect tens of millions of patients in the United States alone, and there are limited therapeutic options available for their treatment.

(2) Based upon this evaluation, we identify what we believe are appropriate and interesting compounds to treat these conditions — compounds that have previously been in clinical trials for other indications. By targeting compounds that have already been tested in humans, we have human safety information about the drug and how it is absorbed after having been taken orally. This information greatly reduces the risk of failure in development.

(3) This approach has allowed us to identify multiple development candidates. In a two-year period since starting the company, we have

identified and initiated three clinical development programs. We expect to initiate additional programs over the next few years. Having numerous shots on goal further reduces our risk, because the future success of the company does not depend on the outcome of a single program. We strive to keep building our portfolio by looking at new products and having more than one option in development at any one time.

Who Has What Role in Which Deal

When you have an understanding of the industry and of the deal processes, you have a fairly good sense of who will play what role in the deal. For example, we participated in a research collaboration with a division of Johnson & Johnson. They came to us for our pharmacology and neurobiology expertise. We highlighted these company attributes in the deal outline. If you don't know what the value drivers are for a particular company, you need to ask them.

When it comes to putting deal terms together, you have to understand what the other side's budget constraints are. It may be important to have a more substantial upfront payment if the deal is being done at the end of the fiscal year and your potential partner has money remaining in the budget that needs to be spent before it goes away. Sometimes, the opposite may be true and back-ending the economics may be important. Adding a premium onto research support may work better to push more of the payments to out years. You have to ask a lot of questions to get the information and details you need to put together an effective term sheet.

Plan a Great Deal

In our company, we have scientists, clinicians, and businesspeople who are part of our business development approach. When we look at new opportunities, we assess them from all angles: business, legal, intellectual property, scientific, and clinical. We look to the people from within those disciplines to offer insight as to what issues we might encounter when bringing a new product forward.

When you do a deal, try to ensure that it will work for both sides; that way, there is less likelihood that the deal will fall apart down the road. Make certain you tell the other side what is most important to you and ask them what is most important to them. You want to be in the best possible position to find mutually acceptable solutions. It's also important to emphasize that you understand the other company will not do a deal it cannot benefit from economically; and they should understand the same about you.

Biotech deals require a great deal of knowledge, understanding, and patience. With the right combination of all three, you can structure a deal that can advance both your company and your long-term goals.

Dr. Lee R. Brettman is co-founder, president, and chief executive officer of Dynogen Pharmaceuticals, Inc. Dr. Brettman was previously an entrepreneur in residence at Oxford Bioscience Partners. Prior to Oxford, he was chief medical officer and senior vice president of medical and regulatory affairs at Millennium Pharmaceuticals, and he filled the same role at LeukoSite, Inc. before the company merged with Millennium. At both LeukoSite and Millennium, he was responsible for building and directing clinical research and regulatory affairs, as well as directing development programs in oncology, GI, and inflammation, including the successful development and registration of Millennium's first marketed therapeutic product, Campath. Prior to joining LeukoSite, Dr. Brettman worked at Vertex Pharmaceuticals, the Schering Plough Research Institute, and the Robert Wood Johnson Pharmaceutical Research Institute of Johnson & Johnson. Dr. Brettman has played a major role in the successful development leading to commercialization of small molecules and biologics including ofloxacin (Floxin), l-ofloxacin (Levoquin), isepamicin, Campath, and Velcade.

Dr. Brettman is a graduate of MIT and received his M.D. from the Baylor College of Medicine. He did his training in internal medicine at University Hospitals of the Case Western Reserve School of Medicine and his fellowship in infectious diseases and immunology at Bellevue Hospital and the New York University Medical School. He is board certified in internal medicine and infectious diseases, and he is a fellow of the American College of Physicians and the Infectious Disease Society of America, as well as a past president of the Connecticut Infectious Disease Society.

Building Successful Biotech Deals

Manish N. Chapekar
General Partner
Montreux Equity Partners

ASPATORE BOOKS

C-LEVEL BUSINESS INTELLIGENCE

Firm's Background and Goal

Montreux Equity Partners is a traditional venture capital firm based in Menlo Park, California, that invests in new and emerging life science and health care companies. We are currently investing out of funds with over $170 million in capital, and we invest at all stages of the company lifecycle, investing in early, intermediate, and later stage private companies.

Most of the companies we invest in are either developing or marketing a therapeutic product. A therapeutic product can be a medical device, a pharmaceutical drug, or a biopharmaceutical drug that can be used to treat diseases. We don't tend to invest in other types of life sciences companies, such as companies that are building research tools or diagnostics that other companies can then use to build a therapeutic product.

Although the companies we invest in are small in size, with eight to fifty employees, these companies are led by very senior management teams. It is very rare that we invest in an idea on paper or an engineer with the product concept. As drug and device development is a complicated business, we need seasoned executives or entrepreneurs who have built therapeutic products before, taken these products through clinical development and regulatory approval, and launched these products in the marketplace.

Our job is to deliver superior financial returns to our limited partners, and hence to identify and invest in private life sciences companies that have the best potential for doing so. There is an unambiguous way to gauge our success in our industry, and that is to deliver superior returns on our portfolio and to do so in a sustainable manner over a long period of time.

Although the objectives are clear, it is hard to distill down into a formula the key factors that make for success in an investment. Different venture capital firms have different strategies that have delivered success. Our personal strategy has been to invest in great clinical therapeutic products and great management teams.

Personal Strategy for Successful Deals

My personal strategy has been to invest in great management teams building great clinical therapeutic products, in that particular order. Average people with a great product will still have average results in our business; great people with average products can still make a good business. We like to invest in management teams that have taken drugs through the clinical and regulatory approval process in the particular specialty and/or have marketed such specialty products and bring a single-minded focus to doing so. We also look to see if the senior teams embrace the concept of "capital efficiency" in executing the plans and if they are acutely focused on shareholder value creation.

On the product side, we like clinical therapeutic products that address large and significant therapeutic markets, have significant human safety data, and have a good intellectual property position. Further, we like products that have low clinical development risk, and a clear, addressable, and relatively short path through clinical and/or market development. We especially like companies developing novel applications of "old" drugs — drugs that have a large and significant human safety experience, either due to their use in a different country, a different indication, a different formulation, or through large safety studies. Given the current sensitivity of the FDA around the safety of pharmaceutical products, it is our belief that it will become increasingly more expensive and time-consuming to prove the safety of "new" molecules and drugs. And finally, we like companies developing products for the specialty therapeutic markets where the products target a specialty group of physicians that a small company can address with a small specialty sales force.

Deal Due Diligence

Our investment process begins with a thorough understanding of a candidate company's technology, products, and the markets in which it wishes to participate, as well as a diligent examination of the backgrounds and talents of the entrepreneurial and management team. While reviewing investment opportunities, we conduct extensive due diligence on a variety of issues:

Market

How big is the market? Is it stagnant or growing? What is nature of the unmet need? What are the competitive dynamics, both today and tomorrow? How would the company position itself against its competitors? Is it addressable with the company resources? Would there be reimbursement for the product? Who is the target physician and patient audience? What is the sales process and cycle? What is price elasticity and earnings potential?

Products

Do we have a good understanding of the safety and efficacy of the product? Does it have good intellectual property protection? How good is the early preclinical and clinical data? What is the indication being pursued for regulatory approval? How hard is it to get products approved for such an indication? Is the clinical development plan clear and addressable? What are the time, path, and resources to clinical approval?

Plan

Is the plan realistic? What are the risks and sensitivities in the plan? How much capital is needed to achieve the plan? Does the business model make sense? How attractive is the business model? Does the company have the talent and resources to execute on the plan?

People

Does the senior management team have the knowledge and experience? Do they bring a single-minded focus to building the products and the company? Can they hire well? Do they embrace the concept of "capital efficiency" in executing the plans? Are they focused on developing the company and shareholder value?

Price

What are the capital and financing requirements of the company for product approval, product launch, and profitability? Can the company be financed in the future? What are different alternatives to liquidity? What might the valuation of the company be at future time points? What are the comparables?

Of course, the objective of this exercise is to assess the risks in the investment to the best of our abilities. We operate in an environment of incomplete and unavailable information. We engage the management teams in presenting the opportunity and addressing these issues for us. We seek feedback from a variety of experts — physicians, regulatory consultants, and buyers — to help us understand these issues and risks. Our final assessment has to do with whether we can clearly identify and assess the major risks and assess the management's ability to deal with those risks.

Risks in Biotech Investments

There are several differences between biotech deals and deals in other industries. First, product development in the pharmaceutical and medical device industry is a capital-intensive and lengthy process. The Tufts Center for the Study of Drug Development estimates that the cost to develop a prescription drug is about $800 million. Additionally, they estimate that it takes between ten and fifteen years to develop a new prescription drug. These estimates are for large pharmaceutical companies developing new prescription drugs for large multibillion-dollar markets. Most of our drug companies are substantially more capital-efficient, but they still need anywhere from $30 to $100 million to develop a drug, and that development still takes anywhere from three to six years.

Second, most of the products have to go through a rigorous regulatory process to prove the safety and efficacy of the drug/device. Before a company's products enter human testing, the company needs to conduct extensive laboratory and animal studies to understand the biological activity, pharmacokinetics, and toxicity of the drug. Then, the product has to pass through a series of human clinical trials (Phase I, II, and III) to systematically understand the safety and efficacy of the drug in humans. Each phase involves a progressively larger population of subjects. Each of these trials can point to a fatal flaw in the product (like safety issues) that would not allow it to proceed further.

Third, most of the products need to be approved for reimbursement by government programs like Medicare and Medicaid, and the private insurance sectors. No matter how attractive the product, if it is not approved for reimbursement, it is unlikely to succeed. Hence, investment in product development and marketing in the biotech/pharmaceutical/medical device industry is an expensive, time-consuming, and high-risk proposition.

Making Investments: Negotiation Approach

I don't like the word "negotiation" when it comes to investments. What we try to do is bring like-minded people together to finance a company so it can move to the next stage of building a product and a company. This does not involve a negotiation in the classic sense of the term; usually when you negotiate the terms of a deal, you execute the deal and then move on, typically to never see that person again. In our case, we negotiate an investment and then work with that person every day for the next five to seven years. During that time, there will be many other negotiations involving financing, acquisitions, and other business decisions. It is a repeat interaction and a collaborative exercise each time. Hence, getting the best deal the first time you make an investment is not what makes a successful investment.

The most important element to a successful deal, in my mind, is a good fit between the venture firm and the management. The second is whether these two groups are on the same page with respect to how to build a company and what it takes to build one. The third step is to agree upon the financing and the terms. Typically, we like to negotiate simple term sheets with the management teams, with a defined share price and a clear definition of how much we put in and how much of the company we get.

Roles in a Deal

Creating a successful deal is a collaborative exercise between the entrepreneur/chief executive officer and the investor. Both the investor and the entrepreneur come to the table with the same objective — to build great products and companies. A great entrepreneur brings incredible passion, commitment, and vision to this common goal. A successful entrepreneur also knows he or she has to work with the shareholders in building these products and companies, and hence must embrace the concept of "capital efficiency" in executing the plans and must be acutely focused on shareholder value creation. A good investor, on the other hand, needs to be aware and understanding of the passion and ambitions of the entrepreneur, and must work collaboratively with the senior management teams to maintain the focus both through progress and the tough times.

Locating Opportunities

There are several sources we use to find investment opportunities. It's important to understand that we, as do most venture firms, go through a multiplicity of business plans, ideas, and companies every year to find a few investments that pique our interest. In a year, we would review about 800 companies, perform extensive due diligence on sixty to seventy new companies, and end up investing in six to eight companies. In looking for new investments, we look for a great investment as well as a good fit for our venture firm.

We receive investment leads from multiple sources. In my experience, the best sources of leads are the management teams we have invested in. Each chief executive officer and vice president in our portfolio companies is a source of unique insights and ideas. These leaders tend to be the richest sources for ideas on investment opportunities. Another good source is our fellow venture capitalists with whom we have co-invested in the past, and with whom we share a common approach to reviewing investment opportunities.

Additional sources we have found useful are physicians and clinical thought leaders in the specialty of interest. These specialists are acutely aware of the unmet clinical needs and the emerging drug opportunities. We meet with them to see what they look at and are involved in. We attend conferences and we research medical journals, industry magazines, and other scientific publications. We are also very fortunate in that many of these thought leaders are the clinicians and researchers at some of the top medical schools in the Bay Area, like Stanford and the University of California in San Francisco.

Another approach I continue to find useful is the old-fashioned cold call. Today we have a great deal of information transparency on early-stage companies and programs. If we find a company that is of interest, we try to get to know the company way ahead of the financing, and we are able to monitor the progress over a period of time before we actually make the investment.

Managing Risk

Venture capital is a form of "risk capital." In other words, there is a great deal of risk involved in early-stage companies that are involved in building and marketing therapeutic products. The only way for venture investors to make great returns is to take some risks. The challenge is in assessing them accurately and managing them to the best of our abilities. Venture capitalists differ in the kinds of risks each is willing to take.

In general, we have been unwilling to take on people risks. Since drug development is a complex, lengthy, and capital-intensive process, we like to invest in management teams that have taken drugs through the clinical and regulatory approval process in the particular specialty and/or have marketed such a specialty product.

We do, however, take on some amount of product risk. The way we manage risk on the product side is by investing in clinical therapeutic products with significant human safety data and a good intellectual property position. Further, we like products that have low clinical development risk, a clear, addressable and relatively short path through clinical and/or market development. Hence, we especially like companies developing novel applications of "old" drugs — drugs that have a large and significant human safety experience either due to their use in a different country, a different indication, a different formulation or through large safety studies. We also manage the clinical execution and regulatory risk to deploying the best clinical and regulatory teams for that particular specialty in designing and executing clinical trials and seeking regulatory approval.

We generally do not take on much market risk. There is lot of information available on therapeutic markets. You generally know what the market needs, how many patients there are and what they suffer from, and who the competition is and how they are progressing. Once the products are approved, the companies do have to take on the reimbursement risk and sales execution risk of appropriately launching and marketing the products.

We also make sure we structure the financing for the company to meet certain milestones that de-risk some critical elements of the investment. We might finance the company to clinical milestones or sales milestones. Sometimes, we might stage the financing so a certain amount of money is invested up front, and upon achievement of certain milestones, additional money is invested.

Mr. Manish Chapekar is a general partner and managing member of Montreux Equity Partners. Mr. Chapekar joined Montreux Equity Partners in 2001 and has extensive experience in new business development and private equity in the life sciences industry. He serves on the boards of Kanisa Pharmaceuticals and Percutaneous Systems, and he has been actively involved with several Montreux portfolio companies, including Coalescent Surgical (sold to Medtronic in 2004), Integrated Biosystems (sold to STEDIM S.A. in 2004), Pulmonx, Skinmedica, Somaxon Pharmaceuticals, and TransOral Pharmaceuticals.

Prior to joining Montreux, he was a manager of strategy and new business development at Pharmacia Corporation (now Pfizer and previously Monsanto Company) responsible for strategic development, in-licensing, acquisitions, and investments in life science companies and technologies. While at Monsanto, he also worked on several growth and strategic initiatives, including developing growth strategies for key products/geographies, accelerating launch of new products, rationalizating the discovery and development processes, and developing new business opportunities through the growth enterprises business incubator.

Mr. Chapekar has received a bachelor's of technology degree from the Indian Institute of Technology in Bombay, a master's of science degree in engineering from the University of Kentucky, where he pursued graduate research in cellular bioengineering, and an M.B.A. from the Stanford Business School.

The Art of the Deal

Daniel Devine
Chief Executive Officer and President
Acceptys, Inc.

Introduction

In this chapter, we will explore some general strategies and observations about the business development and deal-making process with and among biotechnology firms. The discussion will cover a wide range of topics, from conceptual approaches to deal making, to advice on how to optimize your company's business development and deal-making function, to more basic and practical advice on how to assess and document a given transaction. The discussion then moves onto the role of key players in the process and finishes with a more focused look at the special challenges facing smaller companies in the biotechnology industry, which make up a vast percentage of the market in this still-evolving industry.

While the topics covered in this chapter may appear to vary significantly, each carries a uniform message with three major components that are not normally associated with deal making in this incredibly competitive industry. First, meeting the needs of your potential partner in any given deal can be as important as meeting the needs of your own company. In other words, business development professionals should work for both companies involved in a given transaction. Second, some of the major barriers to effective deal making are internal issues associated with the misalignment between business development professionals and their colleagues in operational groups, as opposed to conflicting needs of the two companies involved in the negotiations. Finally, the author of this chapter argues that the financial evaluation of a given transaction, such as the focus on return on investment calculations, is only one factor in determining the attractiveness of the transaction, and it often is not the most important factor.

Developing Win-Win Relationships

In the biotechnology arena, the most notable and important deals involve the creation of long-term relationships between and among the respective parties. For example, the most important and sizable deals relate to the co-development of an early-stage product candidate, and/or the co-marketing of a product closer to commercial approval. These relationships generally span the life of the underlying product, which, if successful, could be over a dozen years long. The same holds true in mergers or acquisitions. When your company is the acquiring entity, the deal doesn't end after the papers are signed — in fact, the hard work is just beginning, and you need the cooperation and participation of the key personnel from the acquired company to make the most of the combination. Even in more transactional deals such as the licensing of a technology or product, the licensor is often in a critical position as the gatekeeper of all data generated before the transaction, and that information must flow effectively over time from one party to the next to aid in the deployment or development of the out-licensed technology or product, respectfully. So, while less important than in longer-term deals, and even with licensing transactions, relationships are very important.

Because of the importance of relationships to biotechnology deals, it is important for each party to treat the transaction as an opportunity to generate a win-win situation. In other words, it is advisable to avoid the typical zero-sum approach to negotiations, under which one party has to win and another party has to lose, and instead focusing on creating a relationship that actually delivers more than either party could achieve alone. By taking this approach, several positive outcomes are possible.

First, by having each party walk away from the negotiating table with a positive view of the relationship, one can greatly increase the chances that the project gets off to the best start possible, because each party is energized and driven by the belief that a good start will help their respective entity. In contrast, if a party feels a relationship is actually detrimental or unfair to its organization, the motivation to make things happen will be lacking or at least lessened.

Second, as with all relationships, the foundation of mutual respect that a win-win approach creates is a great asset to have in long-term relationships where ups and downs are sure to occur. In this industry, the stakes are very high, both financially and on a personal career level, and as a result, strong disagreements among researchers and marketers in a single organization are the norm, nevermind across two independent companies. In other words, the question isn't whether the relationship will become strained, but instead how often disagreements will arise, and how significant those disagreements become. At times, your partner will feel like your most threatening competitor. With a solid foundation of trust created from the very beginning, the chances that a relationship will thrive are increased substantially, despite these inevitable conflicts.

Third, to ensure long-term success and motivation on all parts, the economic terms must be fair to each party not only in the short run, but also in the longer run where the relationship is active and ongoing. For example, if you are negotiating a co-promotion agreement, or a co-development relationship, it is in your best interest to make sure the economic incentives available to your partner are adequate over the life of the deal when the participation of the partner is needed over the life of the agreement. This industry is under a great deal of pressure to perform financially. As a result, for example, no matter how high the upfront payment on a particular deal, once received and booked, your partner still needs an ongoing economic reason to continue to work hard in the relationship. That doesn't mean the partner doesn't want to stay committed, or isn't appreciative of what it received in the past. What it means is that your partner will always be facing a decision about where to focus resources, and you want to make sure you don't lose out to competing projects.

Finally, another reason to find win-win results in negotiating deals has to do with the industry's perception of your company as a partner. Most industry experts have concluded that partnerships, not internal development projects, will differentiate the winners from the losers in the ever-competitive drug industry. Your ability to partner is as important as your ability to do research and development, or to manufacture, or to market and sell. This is already true as evidenced by the Lipitor co-promotion agreement between Pfizer and Warner-Lambert that had a bigger impact on both companies than any other factor. The same can be seen in relationships formed by the leading biotechnology firms. Thus, your ability to be chosen as a partner is critically important for long-term success and value creation. So when negotiating and managing relationships now, you must consider the impact of a failure, or a success, on your ability to compete for partnerships in the future.

In sum, despite contrary practices in the business development fields, a win-win situation is the ideal result, because you are creating a long-term relationship. Think of it as a romance with bigger dollars at stake — if you go into a relationship and your primary goal is to maximize the benefit you get from the relationship at the cost of the other person, how long is that relationship going to last and how happy is your partner going to be? Also, how happy are you going to be when you are "wed" to an unhappy partner?

Developing a Clear Understanding of Your Company's Needs

Before creating his or her own business development initiatives, or responding to a transaction proposed by a third party, a business development professional must understand his or her own company's needs. Like many things in the corporate environment, this is a lot easier than it sounds. What barriers exist that prevent a professional from determining his or her company's needs on an affirmative and not reactive basis?

First, many business development groups are understaffed (after all, such groups are often viewed as cost centers, not profit generators). Also, companies are often flooded with unsolicited business proposals. As a result, group members have little to no time to spend on studying a company's status (which is constantly changing) and affirmatively determining important corporate needs or weaknesses that could be addressed via business development. Ask any business professional from a large- or medium-sized company what percentage of their time is spent on unsolicited transaction proposals versus time spent studying their own company and developing affirmative business development strategies. The answer may surprise you. This shouldn't be a surprise — one of the hardest things for any group to do is to set aside time to plan affirmatively as opposed to being reactive. And you cannot minimize the damage this can cause. In an ideal world, nearly all business development activities would be focused on opportunities first identified by a given company's employees. After all, who knows more about what a company needs than its own employees? But factors prevent this from happening in regular and methodical way.

Another impediment to gaining an understanding of what your company needs is the fact that often no one tells you what the company needs. Manufacturing may have a need, or drug development may have a wish list, and the same holds true for marketing and sales. But only the business development colleague, with the help of senior executives, is responsible for putting all the pieces together in an effort to find a comprehensive solution that can meet the needs of several operational groups. In essence, the business development professional is responsible for putting all needs into context, and engaging in independent thinking and analysis separate from what is communicated to them from various disjointed groups. This issue is closely related but slightly different from the issue of being overworked and having little time to think strategically, but is different in the sense that you must realize some of the best business development solutions will have to come from you, and not the operational groups.

An additional impediment to gaining an understanding of what your company needs has to do with the misalignment between business development colleagues and other key operational and corporate groups. There are structural and practical difficulties that prevent effective communication, cooperation, and teamwork between and among disparate groups. (See discussion below entitled "Creating an Effective Business Development Team.") No person is an island, and without the open cooperation of other groups, the business development colleague's job gets that much harder, particularly in the very complicated field of biotechnology, where cooperation across groups is absolutely essential.

Another weakness in this area has to do with short-term and long-term perspectives. Very often, most employees are focused on this month or this year — this is especially true as you get deeper and deeper into operations. In contrast, the business development professional must take a longer-term perspective, because most of the relationships that result from his or her business development activities will be long-term relationships, which often produce more dramatic results down the road.

Even if a business colleague can form an educated view of corporate needs, and even if such a view is accurate, another barrier to creating a win-win transaction has to do with gaining a consensus within an organization. Too often, colleagues from the same company are found arguing business points in front of a third party during negotiations on a given deal. This problem stems from the fact that having team meetings before a transaction negotiation begins is very time-consuming, and sometimes exhausting. Often, such meetings never take place. And even when such meetings are held, the participants may not be interested enough or prepared enough to have a discussion as to what the corporate needs may be. As a result, team members enter negotiations with different priorities, sensitivities, goals, and expectations.

Can a business colleague overcome these barriers and form a good understanding of corporate needs that is consistent with his or her colleagues and senior managers? The answer is yes, most of the time. (Nothing in corporate life is perfect.) Here are some steps to follow.

First, one must be very disciplined and set aside time to study corporate needs in an affirmative way, and you must solicit direct input from senior executives and operational groups. If you find yourself always responding to proposals, and not forming your own business development ideas and goals, you are not doing your complete job, and you are probably missing the most important parts. You need to work on this with your supervisor, because without his or her support, setting such time aside will be very difficult. You may need additional staff to achieve this goal, but it is money well spent, because business development is the life blood of all biotechnology and pharmaceutical companies that aim to be sustainable enterprises.

An additional helpful step is to overcome and address the misalignment between business development and operational groups. (See discussion below.)

While difficult, you must also have productive and open discussions with your team before meeting with a third party for any given deal. Participants must have the opportunity to prepare for such meetings, and must be allowed to air differences and disagreements. You will probably need more than one such meeting (e.g., the first meeting could be informational, and a follow-up meeting can be focused on forming a consensus). It is always permissible to disagree internally, but to take that disagreement to the outside world can be very destructive.

Finally, it is the business development professional's responsibility to balance short-, medium-, and long-term corporate goals and/or impacts in any given transaction. You will have to make tradeoffs, which is appropriate and often necessary, but you must do so when you are fully aware of what is being traded off. And you only get to that place by understanding the impacts of a given deal over the life of the deal.

Creating an Effective Business Development Team

A business development professional[1] can produce a win-win relationship from a given transaction only if he or she can effectively manage a team of colleagues from all key functions involved in the transaction or the third party development or promotion relationship it creates. This may sound obvious and simple, but this is often far from the case. Why? There are the obvious answers that have to do with the personal capabilities of the business development professional, such as negotiation skills, communication skills, personality, hard work, leadership capabilities, organizational skills, and intelligence, which won't be addressed here.

What may be less obvious is that a major barrier to the business development professional's ability to create a winning transaction has to do with corporate structures. In other words, a good percentage of misalignment between what a company needs and what a business development professional aims for or can achieve from a given transaction has to do with the fact that business development personnel are often in different departments than the key personnel from sales, marketing, manufacturing, and research and development, who are often all necessary to generate value from a given transaction and resulting relationship.

---

[1] The term "business development professional" refers to any person playing a key role in managing the business development, licensing, and/or merger and acquisition functions of a given company, so it may include colleagues with these specific functions, or even a chief executive officer where such chief executive officer is playing an active leadership role in the transaction, which is often the case in the biotechnology industry.

A misaligned corporate structure can prevent the business development colleague from gaining a clear understanding of corporate needs in a number of ways. First, the physical separation of business development managers from operational groups can lead to difficulty in scheduling meetings with operational colleagues located on different ends of a corporate campus, or even different states of countries.

Second, business development groups are generally measured by how a deal looks upon the closing of the transaction, whereas operational personnel are never measured on their contributions to a given transaction (i.e., their contributions are hard to quantify, or the performance reviews of the colleague focus on their day-to-day operational responsibilities). Even when measured, operational colleagues may not get evaluated on a given transaction until the underlying project is rolled out, which may take years. For example, a business development colleague may be rewarded using shorter-term measures like upfront fees, whereas a colleague in manufacturing won't be measured for years to come as production methodologies are developed and optimized. These different and sometimes contrasting performance standards lead to sometimes conflicting goals and varying levels of interest in becoming involved in a given transaction, which often lead directly to poor results.

A third major contributor to the misalignment between business development professionals and their operational colleagues is directly attributable to the fact that many organizations refuse to grant the business development professional enough status to influence operational colleagues. For example, a cross-functional team working on a given co-development and co-promotion agreement may consist of representatives from business development, marketing, sales, manufacturing, law, and finance. However, it is often the case that in such circumstances, senior executives often refuse to appoint the business development colleague as the leader of a transaction team, even though they are generally the best candidate due to their broader corporate perspective, transactional experience, and related capabilities. (This refusal often has to do with the political difficulty of appointing one corporate group as the leader of other groups.) So you have a battle for leadership in many of these situations when the members would be better off focusing their collective energies on the transaction itself. In addition, even in circumstances when the business development colleague is formally announced as the transaction team leader, on a practical level that colleague has little to no influence on operational colleagues because (a) the operational colleagues are rarely made accountable to the leader, (b) the different team members are measured differently (see discussion above), and (c) operational groups often appoint senior managers (or even the leader of the group) to a given transaction to make sure the group's interests are well represented and protected, which makes leadership and management by the business development colleague difficult.

How can a company overcome the issues associated with effectively managing the type of cross-functional teams that are necessary to gain a win-win situation from a given transaction? There are several straightforward, but not always easy, solutions. First, the company must make it very clear that the given transaction is important to the overall company goals, and that the transaction will have impacts on all relevant operational groups. All involved colleagues must understand why the transaction is important to them and their group if they are going to be motivated to work with colleagues from other groups, and to commit the extra effort needed above and beyond their operational responsibilities. Second, all participants in a given transaction must be responsible for their contribution (or lack thereof) to a given transaction at the time the transaction takes place. This can be best achieved by including other transactional team members in the review process, even if from different departments. Third, the corporation (meaning its senior executives) must make it very clear that operational groups must follow the lead of the business development colleague — that is, they must accept being constructive and value-adding members of a team led by another department. And those that don't follow the lead must be held accountable. That is very difficult to do, but it is the only way you are going to make a vice president of marketing follow the lead of a director from business development. The influence and power of the business development colleague must transcend his or her department name and management level. Finally, a great deal of benefit can be achieved by having business development professionals spend more time with operational colleagues outside the transactional arena. For example, appoint a business development colleague to a marketing project so he or she can learn about the function, but also so camaraderie and trust can be built across the functions. This is another way to move capable and appropriate people from operational units to business development functions. These colleagues can carry a good deal of credibility and respect when dealing with their former colleagues, as compared to a new hire from business school with no operational experience. This only makes sense when the operational colleague can work toward overall corporate needs, even at the expense of their former operational group.

Understanding Your Potential Partner

Perhaps the greatest weakness shared by most business development professionals relates to their understanding of the needs of potential partners. As discussed above, one rarely has enough time to determine the needs of your own company, nevermind a third party. But to achieve a true win-win relationship, this aspect of the deal is absolutely critical.

And the inquiry does not stop at determining what the other party says they are trying to get out of the transaction. Why can't you exclusively rely on what is said by the person on the other side of the table? There are many reasons. First, a good percentage of business development professionals are not that straightforward. In other words, they believe that if they reveal what they really want, it will be used against them as leverage in negotiating a deal. Second, as mentioned above, your counterpart suffers from the same institutional weaknesses that you do, including limitations on their ability to determine what their respective companies need, and difficulties in getting the full cooperation of their operational colleagues. Finally, it is unlikely that your counterpart will know all the things your company can bring to the table in a given transaction.

In order to maximize what your partner gets from a given transaction — which benefits you because the more they get, the more you can get — a business development professional can follow these guidelines:

1. Study your partner carefully. Use the Internet, company Web sites, company literature, and any other source you can access to learn about your partner. One great place to look is in SEC filings and

annual reports. You won't always have time to do this as thoroughly as possible, but do the best you can — it is well worth the effort. This not only gives you a sense of what your partner may benefit from in a given transaction, but during negotiations, your knowledge of their company will become apparent and will add credibility and trust to your relationship with them. Also, as with getting to know your own company, a business development partner should get a sense of the potential partner's needs over the short-, medium-, and longer-term in cases when a resulting relationship would last that long.

2. Don't limit your conversations to those with your business development counterpart. Before and during negotiations, try to engage their complete team, including any operational or corporate colleagues your counterpart brings to the meetings. Ask questions about what they are looking for from the transaction. It may well be the first time they've been asked such questions. You may even consider asking your counterpart to include key operational colleagues in the negotiations or side meetings — that way, you can hear from them directly what they need or want from a given relationship. And offer up the opportunity for the operational groups from each company to meet (assuming there are no downsides to such discussions, and of course waiting until after you've made progress toward an agreement — this has to be managed delicately).
3. Match your partner's needs with what your company can offer in a given transaction, and vice versa. Go through each major function, and ask whether there are any synergies between the companies. For example, assume the transaction in question relates to your company acquiring the rights to market Drug X in Europe from Company A. During your preparation for negotiations, however, you find out that Company A has had some production capacity problems on Drug Y (which isn't part of the deal). You also know, based on your own internal due diligence, that your company has excess capacity. This may be a good opportunity to expand the relationship to include manufacturing of Drug X, freeing up capacity in Company A to make other products, including Drug Y. This also solves your excess capacity problem, and gives you

control over the production of a drug you are marketing. This is just one example of a possibly elegant way of expanding the benefit of a relationship for both parties involved, that would not have been identified if the business development professional wasn't aware of the other company's needs. Approaching deals in this manner, with a win-win focus, can expand the pie that the parties are splitting, to everyone's benefit. It also relieves stress, because the teams are working together more than they are against each other, giving the relationship a solid foundation of trust and positive energy — key assets once the deal is signed and turns to the execution stage.

4. One simple way to summarize the suggested approach is an age-old directive: do your homework before sitting down to negotiate a deal. Have it as your goal to be the most informed person at the table, not just about your company, but about the other company as well and how the two can work together. The more work you do, the more knowledge you will have, and the more ability and influence you will have to leverage in order to create win-win solutions. If you are ignorant, you are not going to be able to be creative.

The Role of Return on Investment: The Importance of Scenario Analysis

In the drug industry, most business development professionals focus on the return on investment (ROI) from a given transaction, and this is appropriate. After all, companies exist to make profits and returns for their shareholders. However, what is less clear is how much weight you should put on ROI in any given transaction or potential relationship. There are several reasons why one should be very careful not to put too much weight on ROI calculations.

First, nearly all ROI calculations involve assumptions around factors that are partially or completely outside the control of either company in a given transaction, and are in all or most cases subjective. Consider a co-development/co-promotion agreement. One has to make assumptions about development expenses needed to get regulatory approval (including clinical size and costs), timing of regulator review, the chance regulatory approval will be granted, market size, market penetration, pricing, cost of goods, current and future competition, and government programs such as those including reimbursement mechanisms. There are many opportunities to make mistakes when an ROI calculation involves such complicated factors. In fact, it is inevitable that mistakes will be made, so you can only help to minimize, not eliminate, mistakes. And given the fact that most deals in this industry are long-term and involve significant dollars, any mistake in a key assumption can dramatically impact your ROI calculation.

In addition to the inherent inaccuracy associated with ROI calculations in the biotechnology and pharmaceutical industry, another reason to be careful not to put too much emphasis on ROI is because doing so can lead to an inappropriate deal structure and focus. For example, if both parties focused primarily on ROI, you would see a good deal of negotiations and focus on upfront payments, or fixed annual payments, since they contribute greatly to ROI by coming in the early years and because no risk is associated with such payments coming due. However, when forming a win-win transaction, these one-time, one-way, and non-contingent payments often don't contribute to a sense of partnership in a relationship, because they don't drive parties to grow the overall profit of a given relationship.

Another reason to be careful about putting too much weight on ROI is the fact that other factors must be considered in any relationship. More specifically, the key question involves capabilities. For example, in drug deals, you must first ask whether this company is going to help your product get to the market and help grow the sales, market share, and profit coming from such drug. If the product doesn't get to market, there is no ROI. Even key items such as royalty rates, or profit splits, must take a back seat to these key questions.

In sum, while ROI calculations are necessary, a business professional would be well advised to avoid spending too much time on ROI calculations at the expense of other key transaction or relationship factors, and putting too much weight on ROI at the expense of other factors.

One proven approach to balancing ROI with other deal factors is to structure a deal that provides continuing and strong financial incentives for both parties to perform over the life of a transaction. So, for example, put in a profit split mechanism that isn't fixed, but is instead focused on future successes and activities. A royalty rate that increases as certain sales milestones are hit is one example, or a higher profit split percentage to a party that makes more sales calls. There are many ways to achieve this goal, all circling back to creating win-win situations.

One way to use ROI constructively is to run various scenarios for any given deal. For example, change your assumptions about royalties, upfront payments, manufacturing costs, sales figures, advertising costs, development costs, and other major items. Run a best case scenario and a worst case scenario. This type of exercise, while demanding a bit of time and energy, is almost always very productive and useful. While each specific scenario has weaknesses, primarily due to underlying assumptions, what the exercise can provide is very useful learning that may hold true across a number of scenarios. For example, it may tell you that generally, you should be concerned more with royalties as opposed to upfront fees or milestones, or that you should aim for a profit split as opposed to a royalty, or that you should develop an escalating royalty based on sales levels. Importantly, this exercise also could result in setting a standard for walking away from the deal if the downside scenario would effectively threaten your company's survival in the absence of safeguards. So while ROI has inherent and significant weaknesses, it can be used productively if used in this context.

The Importance of Data

In the biotechnology field, data is at the heart of almost every major category of deal, including license transactions, co-development and co-promotion agreements, and mergers and acquisitions. The data a company presents must have the following characteristics to facilitate a given transaction.

First, it must come from a credible source(s). Has it been repeated? What are your controls? Has it been validated by a disinterested third party? Are there inconsistencies?

Second, it must be comprehensive and without any major gaps associated with important aspects of the deal. If you are missing key elements, most partners will assume that either you were negligent in not generating the data or in including it in your package, or that you have the data but did not present it because it was damaging to your case.

Third, the data must be understandable — that is, it must be presented in a form that is understandable to third parties. This is a major pitfall in business development. There are two main problems with failing this test. First, most of your business development counterparts are very busy, and get extremely frustrated with disjointed and hard-to-understand data. Very quickly, they will give up and move to another deal. Second, your counterparts will make a mental association between poorly communicated data and poor data. In other words, it is very easy to reach the conclusion that the communication of the data was poor because the data was poor. That is a deal-killer and must be avoided at all costs.

Documenting Your Relationship: A Critical Process

There are few things in the industry that people dislike more than legal documents. Not only are such documents long, cumbersome, difficult if not impossible to read, and complicated, but the drafting process generally happens at the end of a long and tiring negotiation process, when participants are running out of energy and want to put the transaction behind them. This is a common and fundamental mistake.

The documentation process is absolutely critical, as it memorializes the agreement between the parties, and then becomes a living and operational tool as people revisit the document as the relationship matures and as expected and unexpected situations or circumstances arise, many of which can be controversial. More than any other party, it is the business development professional's responsibility to make sure the documentation process is effective.

This can be achieved by first ensuring that all relevant parties review their respective parts of a given agreement. That means manufacturing colleagues must take the time to read, understand, and comment on those sections that deal with production, and the same holds true for marketing, sales, research and development, and so on. Many operational folks are tempted to flee when the major terms are agreed to, so a system and structure must be put in place to ensure that this doesn't happen. And that responsibility needs to be put on the operational colleagues as well, as such colleagues should not have the opportunity two years from now, when they become involved in the relationship, to look back and complain that the document was unreasonable.

The other goal of the documentation process is to provide a fixed structure for how the relationship will be managed over time, and how it will respond to a variety of possible events. Who will be in charge of the joint effort? How often do the parties meet, and where? What happens in the case of a disagreement between or among the parties that can't be solved at the operational level? What happens if we have an unexpected event at a plant that cuts off production? What happens if one of the companies is acquired by a direct competitor? What happens if a party files for bankruptcy? What if your partner in a co-promotion deal violates laws covering proper promotion and advertising? How would a product recall be handled? This involves a good deal of anticipation and forward-looking activities, which is not an easy thing to do at the end of a long road of negotiations. But it is very important to handle these issues up front, because trying to address them when they arise is not optimal.

The Chief Executive Officer's Role in Deal Making

The chief executive officer (CEO), as with the business development professional, must first be focused on the big picture. The following are examples of questions that should be asked and answered by the CEO:

a. How does this transaction impact our core assets?
b. Is the transaction consistent with our overall strategy for growth and profitability?
c. How does the relationship impact the company over the short, medium, and long term?
d. How does the transaction impact our ability to raise funding to support us going forward?

e. Is this the right partner and why?
f. Does this deal prevent us from doing other more important deals or activities?
g. How does this deal lead to shareholder value?
h. For smaller companies, how do I balance our need to do deals with larger companies with our need to protect our key assets?

In addition to making sure the deal fits with the overall corporate goals and needs of the company, the CEO and other senior executives must also be facilitators. This means the CEO must make sure the right internal system is in place to maximize the value from a given deal. This means making sure all key operational groups are involved and responsible for contributing, the business development leader has enough influence to lead the effort and drive the process effectively, and all relevant employees know why the deal is important to the company and how it impacts their respective functions.

Finally, after the transaction is signed, the CEO must be involved in managing the project internally, especially as it relates to the conflicts many deals generate within an organization. More specifically, on many deals there is going to be a segment of your colleagues that think you don't need to do the deal — for example, that you could have developed and/or marketed a lead drug without any help from a partner. Not everybody in the organization is going to initially get on board, so managing that process internally is very important for the CEO as well.

The Chief Financial Officer's Role in Business Transactions

The chief financial officer (CFO) and his or her organization are often responsible for ensuring that the economics of the deal make sense for the company in the short and long term. This analysis must include a situational analysis (e.g., what do the best and worst case scenarios look like and how do they impact the company?). In addition, the CFO's organization must make sure the financial controls and systems are in place to support the relationship and that the financial data generated internally and by your partners is accurate and has integrity. Beyond the traditional roles of the CFO as outlined above, it is critical in the biotechnology industry to get CFOs involved in deals early on, because they can bring a good deal of value well beyond traditional accounting and financial functions. For example, as the CFO knows more about the financial status of the company both now and in the future, his or her input on deal structure could be critical (e.g., Should you put more weight on gaining upfront fees or on royalties down the road?).

The Role of the Chief Scientific Officer in Business Transactions

In the biotechnology industry, the chief scientific officer (CSO) and his or her organization play a key role in every important transaction. It is absolutely necessary for the CSO to be active and engaged. The primary responsibility of the CSO is to provide data to the other side in a way that can be easily read, absorbed, and understood, and evaluate the data coming from the other side.

In deals where assets or technology are leaving a given company, it is not enough for the CSO to be a pure information provider. This is a very technical and complicated field, and even the best presented data is difficult to follow and understand. As a result, the CSO and his or her organization must play an active role in educating the other side, walking them through the data, and ensuring that it is well understood. As mentioned before, when something is hard to understand, people will assume the data is poor. It should be noted that this is not always an easy task, because there is a good chance the CSO or his or her organization doesn't support giving up rights to drugs or technologies developed internally, or there is an equally good chance the CSO or his or her organization feels their job was done when the decision was made to license or sell the asset. One must keep these factors in mind and ensure that they don't operate to sabotage a deal.

In addition, the CSO's organization must play the leading role in evaluating all technologies and drug candidates being brought into their respective company. It isn't enough for the CSO to evaluate the data, but the CSO and his or her organization must be ready, willing, and able to take on the responsibilities called for in the deal. You don't want to hand a drug candidate to an organization unless that organization is fully behind spending resources to develop the candidate. It would be close to a disaster to have your CSO inform you after the deal is done that while the data from the other party was accurate and credible, he or she doesn't believe the deal was in the best interest of the organization. In a related way, the CSO must be responsible for informing the company as to the impacts the deal would have on the other research and development activities of the company. Resources are by definition limited, so any new deal will by definition take resources away from a current internal program, and that impact needs to be understood.

Taking Investors' Interests into Account

The question isn't whether investors or shareholders should be considered when involved in business development activities, but instead how much weight to give such interests, and when to involve them in a particular deal. The answers to those questions vary greatly from case to case, and it isn't productive to go through each one. What is important is for the CEO and other senior executives to be sensitive to what the value drivers are for the outside investor, which are often financially driven and related to the short term, and balance those interests against the potential strategic and often longer-term competing interests of the company. Sometimes, the interests are the same.

Unique Aspects of Business Development in Small Companies

Small companies need to be particularly strategic and focused when it comes to exploring and pursuing business development opportunities, because you don't have the time or resources to do a worldwide search for opportunities, and you can't readily absorb failures as easily as a larger company. This is not as easy as it seems, as there is often overwhelming pressure on small companies to make something happen fast, which seems to work against focus and strategic planning at times – and the belief that deals with other companies are shortcuts to reaching end goals. The temptation to do a deal is much stronger in small companies, because deals can have more dramatic impacts on the small company.

The resolution to this danger is very simple — you must first spend the time necessary to form an accurate list of priorities for your company, and then address each priority starting from the top. It is important to maintain discipline and not move down the list without checking off higher priorities that have been realized. This will give you the greatest chance for success, because the more things you focus on, the higher the probability you will make a mistake or not maximize the return from a transaction, which is something small companies can ill afford.

Another pitfall to avoid as a small company relates to leverage. More specifically, try to engage in business development activities around a key priority when you aren't desperate to achieve your goal. Desperation not only weakens your negotiation position, but it may well scare off potential partners all together. Like most healthy relationships, partners look for strong partners, especially in this industry, and so signs of weakness can be devastating. As with most things in a small company, this is not easy to avoid, because most small companies in this industry are by definition a bit desperate. So how do you address this? That is hard to answer, because it is so company-dependent, but one general rule is to try to negotiate when you have enough time for talks to be extended for several months to a year. This way, you don't communicate to the other side a need to close right away, which is never a good position to be when you are the weaker or smaller company. Another way to avoid this pitfall is to negotiate for key relationships as soon after a funding as possible, so you are healthy from a cash perspective, which is the most important asset of any small company, except perhaps for key products.

Another aspect of business development in a small company is the conflict between a driving need to do deals and the need to protect your core assets, which are often limited. This tension is real, and it faces nearly every small company in the industry. On one hand, there is a great deal of pressure to partner with a more established industry player, because it gives you cash and credibility, which are two very valuable assets for a small company. However, smaller companies generally have one or a few key assets (e.g., drug candidates), and giving up rights to your main assets can have a very dramatic impact on the residual value of your company. This is a very difficult conflict to manage. Every situation is unique, so there are few general rules that apply. That said, it is fair to say that all small companies must be very careful to ask one key question: Despite the short-term benefits of this deal, how does it impact the longer-term sustainability of our company, and what do we have left? The answer may be that you have nothing left, and doing the deal could still be the right thing to do, but it would be a mistake not to ask the question.

Perhaps the most difficult thing for a small company to determine in any negotiation is when to walk away from a deal. As mentioned, deals are critical for small companies for many reasons, including cash generation, building credibility in the market (e.g., external and independent validation), and leveraging the expertise of bigger companies in a way that increases the chance your drug or technology will reach the market. These deal drivers make it very hard for a small company to walk away from a potentially promising deal, which results in small companies being very vulnerable to larger, more established, and more stable companies.

So, how can a company protect itself? The short answer is that you can't, but there are some steps you can follow to minimize the damage that can be done due to such vulnerability:

a. Set up alternatives. Don't put all your eggs in one basket — look around for two, three, or even more alternative transactions with different parties that can give you the value you are seeking from a particular type of business transaction. It is very tempting to focus on one company, especially if it would bring you the cash and credibility you need, but this can be a mistake, because without options you have little to no leverage. On a related note, what you ideally want to do is motivate your partners to act before others come into play. Setting up this type of competitive situation is one of the only ways a small company can move a bigger company faster than the bigger company prefers to move.
b. Do it early. Don't wait until you are running out of cash to start negotiating a deal. In this industry, deals take a great deal of time, generally over one year. So make sure you can survive as long as the negotiations go on.
c. Do careful due diligence on potential partners before sitting down and negotiating. The time and effort necessary to do this type of homework is nothing compared to the lost time and resources

associated with negotiating with the wrong party for six months when you could have ruled that party out up front.

Mr. Daniel Devine is the chief executive officer and president of Acceptys, Inc., a biotechnology company focused on the development of human antibody therapies for cancer and infectious disease indications. Immediately prior to founding Acceptys in 2002, Mr. Devine managed the international business development team for Pfizer Pharmaceuticals. As such, Mr. Devine was responsible for many significant business development activities, including the identification of licensing candidates for key and emerging markets, identifying potential acquisition targets and analyzing those companies from an operational and financial perspective, the divestment of non-value-added mature products, the acquisition of several pharmaceutical businesses in Latin America, and structuring region-wide relationships with contract sales organizations to help support second-tier Pfizer products. Mr. Devine consistently worked with investment banks and other financial institutions supporting Pfizer's businesses, and was active in the negotiation of relationships with third party collaboration and/or license partners.

In addition to leading the business development team, Mr. Devine helped establish and directly managed Pfizer's first centralized international e-business team. In this role, Mr. Devine identified best practices of pharmaceutical companies in the use of the Internet to drive top- and bottom-line growth, and co-developed Pfizer's global e-business strategy, functions, and objectives. Prior to joining Pfizer, Mr. Devine was a key member of the Warner-Lambert/Parke-Davis' CFO organization, and was responsible for substantial changes in the licensing, manufacturing, treasury, and tax planning functions for the pharmaceutical business, all of which led to significant bottom-line savings for the organization. Some of the transactions in which Mr. Devine was heavily involved while at Warner-Lambert included acquisition of Hickson Pharmaceuticals, Jouveinal Pharmaceuticals, and Agouron Pharmaceuticals; licensing transactions with Fujisawa, Pfizer, Menarini, G&M, Sankyo, and other significant pharmaceutical companies; and proposed and actual merger negotiations with American Home Products and Pfizer. Mr. Devine was also the catalyst for Warner-Lambert's establishment of over $1 billion in manufacturing capacity in strategic markets. Earlier in his career, Mr. Devine spent seven years as an attorney at a prestigious Wall Street law firm, focusing on large corporate, tax planning, and litigation matters, including mergers and acquisitions, divestments, and financings.

Mr. Devine is a member of the Biotechnology Industry Organization, the New York Biotechnology Association, and the Biotechnology Council of New Jersey. Mr. Devine earned a degree in economics from Cornell University, was second in his class at the State University at Buffalo Law School, and earned his M.B.A. from Columbia University.

Mergers & Acquisitions, Collaborations, and Partnerships

Mark I. Gruhin
Shareholder, Head of Corporate, High-Tech, and Bio Tech Group, Head of Business, Finance, and Real Estate Group
Schmeltzer, Aptaker & Shepard, PC

Biotech Company Roles

Before even discussing how biotech deals are structured and executed, it is important to first understand the main motivations of key players at the table. Within the biotech company itself, a chief executive officer is trying to grow the business, expand opportunities, and bring success to his or her biotech concepts. Pursuing such interests requires both sources of capital and additional collaboration money. The funds are essential to further the growth of the company, but serve as a proof of purchase at the same time. The chief financial officer, then, is charged with assessing the risks and rewards of any transaction as it relates to a company's balance sheets and cash flows, bringing dollars-and-cents value to any transaction. The final key player within the company is the chief technology officer or the chief scientific officer, who brings a great deal of technical expertise to the particular industry they're representing. In biotechnology, the chief scientific officer has a hand in outlining the company's science structure, while seeking to bring credibility to the equation and marketplace alike.

In terms of outside investors, venture capitalists specifically bring a lot to the table in the sense that they can support a deal with both reputation and money. The role of a venture capitalist may be less important in a collaboration deal, but its credibility can still add value. While with a mergers and acquisitions (M&A) transaction, a venture capitalist's role becomes more involved, as it is likely it controls either the company or the return on its investment.

Value

The greatest assets for a young biotech company are compounds, which have begun to undergo FDA regulatory processes whether they are Phase I or Phase II. Big Pharma is interested in biotech companies that have these compounds either because their pipeline of drugs is drying up, or because they have a failed compound of their own. If a biotech company has compounds that are in Phase II and beyond, buying that compound or business will put a big pharmaceutical company that much closer to producing a commercially viable and marketable drug.

A young company could also be attractive if it is selling a process or an approach to discovering drugs that may have value. Again, it depends upon the needs of the particular buyer, as well as what the seller has to offer. It is appealing for a biotech company to have both the size and credibility to which pharmaceutical companies would be willing to lend their financing and expertise as well.

Valuation models are constantly moving, given the market and the climate, but they are generally based on a multiple of EBIDA, a standard for any industry. Biotech companies, however, have a high burn rate. They are research and development shops, so putting value on a biotech firm can occasionally incorporate more intangibles than in another, more traditional industry. A biotech company is much like a software company in that its intangible assets or value generally lie heavily in intellectual property such as trade secrets and patents. In software, it's patents to software, whereas in biotechnology, it's patents to compounds or processes in discovering those compounds.

The Aspects of a Deal: From Negotiation to Pricing

Once a company has demonstrated value and proven its worth as a viable investment, there is certainly room for negotiation in biotech deals. For example, if the deal involves collaboration on a particular compound, there is often negotiating space in terms of upfront payments, milestone payments, and ownership of the drug as it pertains to future success. Companies may negotiate a royalty stream. These areas may offer room for negotiation, and they can also be the weak points that kill a deal if one party is not willing to go as far as the other wants.

Due Diligence

The process of doing due diligence on a potential biotech partner or M&A candidate is the same as it would be for any deal, and it is generally multifaceted. A company would do legal due diligence, which involves reviewing the corporate documents, the corporate governance, and all aspects of how the corporation was financed.

Intellectual property due diligence can include an evaluation by patent counsel or other patent expertise, as well as an evaluation of the ownership, processes, trade secrets, or intellectual property possessed by the company. Additionally, other due diligence will include exploring relationships with third parties, strategic relationships, and discussing those relationships with those third parties. Red flags in biotech deals generally stem from problems with the technology or ownership. They can also pop up because of issues with the financing and structuring of the company, potential litigation, or other risks common to any business deal.

Pricing a Deal

Certain factors, such as a public versus private company, can impact a purchase price in a biotech deal. Sarbanes-Oxley has added a tremendous amount of cost to public companies, adding to operating costs, already higher than those of private companies. The benefit of a public company, however, is that its paper ("corporate stock") is liquid. The company will often obtain a greater multiple on its paper because of this liquidity, so it has greater potential to use this value in acquiring other companies or assets than a private company would. There are many factors that can affect pricing, including but not limited to the expectation of success of a company's compound or process.

The Threat of Litigation

One never knows where a piece of litigation is going to come from, and obviously that's going to vary with regard to a private company and a public company, but with a public biotech company, there exists increased risks relating to public companies and dealing with public markets, particularly in this era of Sarbanes-Oxley. In a private company, however, there are many kinds of litigation that might arise, but they are less likely to involve investors, other than disagreements with venture capitalists. For example, no class-action security lawsuits are likely to come up.

Breached Representations

When formalizing a deal, companies will make a series of representations involving every aspect of the business, and if they later breach those representations, problems will arise. Litigation generally ensues when a company pays too much for a deal and the shareholders suffer. It can also develop when a purchaser finds itself in unexpected litigation that a seller anticipated prior to the transaction but did not adequately disclose. If in fact a company doesn't do what it represented it would, or represented what it planned to do inappropriately, it becomes vulnerable.

Creating Partnerships

In a partnership or contractual relationship, what defines the terms of the relationship is the contract.

A merger and acquisition is more complicated, as it is like a marriage. It is combining two businesses into one, so there is some difficulty in that, particularly in terms of corporate culture. Compatibility is important, so they should have an appropriate blend of assets and human resources. Often, an overlap occurs when the two company cultures are merged, so there is an elimination of expenses, whether it's people or assets, which can sometimes be painful.

The most beneficial partnerships are ones in which each business can get what it needs when it needs it. For example, a young biotech company may be less concerned with royalties, because it is in greater need of immediate cash to continue its existence. This does not mean that in time, it wouldn't be interested in royalties, but from the time they enter into a collaboration to the time it becomes a commercial product, biotechs can come and go before royalties even become an issue. As is the case in any business, the companies involved are going to try to exploit both sides of the deal to get as much as they possibly can. The biotech's ability to leverage depends upon the interests of the buyer as well as the interests of other potential buyers. If they have a product that's potentially competitive in the marketplace, their leverage is better. If they're competing against thirty or forty other different companies with comparable or similar types of products, they won't be able to extract as much.

When one company buys another during an M&A transaction, the question is always whether they can get to the ultimate goal — in this case, bringing commercial compounds to the marketplace — through organic, internal growth, or if they will be able to move more quickly by acquiring a company. If they decide to do the latter, it just comes down to a matter of valuation and how much they are willing to pay for a deal.

Key Advice for Young Biotechs

To gain the attention of Big Pharma, a young biotech company needs to possess something that will feed the potential buyer's needs. This may be a quicker and better way to find drugs, or patents on particular drugs they have interest in bringing to market. This might also include other sorts of approaches to dealing with current drugs or discovering new ones. Many young biotech companies become a one-horse town, having discovered a compound that may have failed as they went through the processes, and are at a loss as to where to go next. Having something that is going to add a lot more depth to a company might be more attractive than just trying to hit a homerun on one compound. Whatever the case, because of the complexity, high failure rate, and high burn rate in successfully discovering new medicines, a young biotech better be prepared to have strong financial supporters in order for it to survive the industry it is in.

Mr. Mark I. Gruhin is a shareholder at SAS, and he heads both the corporate, high-tech, biotech group and the business, finance, and real estate law sections.

He is a general corporate lawyer with extensive experience representing domestic and foreign corporations, partnerships, limited liability companies, nonprofits, ESOPs, and cooperatives in their formation, day-to-day operations, and corporate and financial transactions.

In addition, Mr. Gruhin has experience representing national lenders in secured financing involving real estate, stock pledges, warrants, and other collateral. His corporate experience includes structuring transactions, drafting and negotiating various transactional documents in connection with asset sales and acquisitions, stock acquisitions, mergers, private placements, venture capital transactions and lending transactions, and conducting due diligence in connection with such transactions. As general corporate counsel, Mr. Gruhin advises clients on a variety of corporate, human resource, and litigation matters that arise in the ordinary course of business.

Mr. Gruhin has been recognized as a leading Internet lawyer and is identified as a "Top Ten Internet Lawyer" in Inside the Minds: Internet Lawyers *(Aspatore, 2001). In that regard, Mr. Gruhin was also a co-founder of a software Internet company whose business is to provide software and service to companies in managing their hiring process over the Internet. In the biotechnology area, Mr. Gruhin has been recognized for "Deal of the Year, Avalon Pharmaceuticals, April 2002" by* Washington Business Forward Magazine.

Mr. Gruhin has been featured in "Ask the Attorney" columns in Red Herring *and* Upside *magazines and has written for the* Entrepreneur Support Series *of the Michael D. Dingman Center for Entrepreneurship at the University of Maryland.*

Strategies for Successful Deals

James T. Gunton
General Partner
NJTC Venture Fund

Introduction

NJTC Venture Fund is an $80 million venture fund investing in all industries. In 2001, the fund began investing primarily in New Jersey, and as of mid-2005, it has invested in more early-stage New Jersey-based companies than any other venture investor. Nearly 50 percent of NJTC Venture Fund's investments are in the health care category.

Strategy for Successful Deals

NJTC Venture Fund's objective is to build businesses that positively affect their industries. As a high-risk investor, NJTC invests in companies with the potential to obtain returns of ten times the investment.

Creating a successful deal is highly correlated with the management team. Having founders possessing the tenacity to adapt to the difficulties that arise is critical to creating a successful business. A receptive market is also required to consummate a successful deal. Prospective customers must need the product and lack a viable alternative.

Negotiating a Deal

NJTC Venture Fund's negotiation approach is to first ask the entrepreneur what he or she wants from the deal. After confirming the entrepreneur and prospective investor are not too far apart with regard to terms and valuation, then it is possible to close the gap. If the two parties are initially too far apart, it is usually better to agree to disagree and move on.

Generally, it works best if the entrepreneur's objectives are in line with the investor's. Is the entrepreneur building a business to change the world, to make money, or for personal reasons? Answering these questions as early on in the dialog as possible can simplify the negotiation.

Finding New Opportunities for Deals

NJTC Venture Fund is a seed and startup investor that identifies investment opportunities in part by working with universities and corporate research groups in its targeted New Jersey market. Partnerships are affected with NJIT, Princeton University, Rowan University, Rutgers University, and Stevens Institute of Technology, among others. Innovation is found among the students and faculty.

In addition, the pharmaceutical and telecommunications industries in the area spawn startups founded by experienced individuals knowledgeable about their sectors. New Jersey is known as the country's "Medicine Chest," with approximately two-thirds of the world's pharmaceutical companies present in the state. Verizon, Lucent, AT&T, and Telcordia (the former Bell Core) are also located in New Jersey.

Mistakes in Deals

A common mistake for an entrepreneur, when making a deal, is to focus on valuation instead of the deal's terms. If the prospective investor proposes the company should be valued at $10 million, most entrepreneurs tend to focus on that figure and view it as a report card. The entrepreneur thinks, "XYZ Corporation was valued at $20 million, so my business should be worth more."

The entrepreneur should instead view valuation in the context of the entire set of terms. Upon a sale of the business, the entrepreneur may end up getting more because he or she negotiated superior terms — such as an attractive liquidation preference — despite a smaller valuation.

Another common mistake is not hiring experienced professional service providers. A savvy entrepreneur will surround himself or herself with professional service providers (e.g., lawyers, accountants, or commercial bankers) who have built their practices around the special needs of emerging growth companies. The deleterious effects of service providers hired out of the Yellow Pages are quickly discernable.

Sharing Advice

Advice shared with venture capitalists just entering the business is to avoid focusing on one opportunity at the expense of developing other potential deals. By working on several opportunities concurrently, an investor does not feel as obligated to a single deal in the face of mediocre due diligence reports.

M&A and Partnership

Evaluating Partnerships

A common target customer can be an effective basis for partnership. For example, Valera is a specialty pharmaceutical vendor that sells to the urologist community. The company develops a hydrogel implant for patients with prostate cancer. Valera would have a good basis for synergy in partnering with another company that sells complementary products to the urology community.

Technology that augments the value of another firm's franchise can also represent the foundation for a successful partnership. For example, Enzon increased its market capitalization to in excess of $1 billion by augmenting Schering-Plough's existing product line through a technology called "pegylation." The partnership was attractive because Schering-Plough didn't have to change very much — pegylation simply extended the value of their existing products.

The Role of Investors and Executives in a Deal

An investor's role in an emerging growth company can be multifaceted. An experienced investor provides guidance on whether a contemplated business transaction is competitive relative to market terms. An active investor will also introduce relevant business partners to its portfolio investments. Finally, an investor can promote the solvency of the entrepreneurial business to large prospective partners concerned about the staying power of a startup.

The chief financial officer tends to be more involved with the nuts and bolts of the deals. For example, in anticipation of a partnership with a big pharmaceutical firm, one chief financial officer converted from a C-corporation to an LLC legal entity, because she correctly recognized that an

IPO for the business was unlikely to occur. An LLC structure optimized shareholder value.

The chief executive officer must deliver a vision compelling enough to attract investors, partners, and recruits. In addition, the chief executive officer is responsible for the company's strategic planning.

Due Diligence

One aspect of a partnership to consider during due diligence is whether the big pharmaceutical company is truly interested in commercializing the startup's technology. A partnership occasionally is prompted by the larger vendor's desire to protect its franchise by freezing competitive alternatives. Specifying a minimum guaranteed sales level or rate of expenditure can help protect against this abuse.

Valuation

There are several different ways an investor will consider valuation. The first examines the size of the market: the larger the market opportunity, the larger the potential valuation. The more saturated by competition the market, the less you can argue for a high value unless you are clearly unique or superior.

Comparables are another way to consider valuation. A review of the public market where companies in similar stages of drug development trade, or of recent public acquisitions, can be informative.

A return model is typically created by an investor to quantify the projected return of investment. "Pessimistic," "Expected," and "Optimistic" scenarios are modeled, and a blended return is computed. Assumptions are incorporated about how much capital will be required and what an acquirer will eventually pay.

Conclusive Comments

We are fortunate in New Jersey to have extensive discovery and development activity in many therapeutic areas. New Jersey was the first state in the country to create a stem cell institute; a tissue bank also exists locally, at the Coriell Institute. Rutgers University, University of Medicine and Dentistry–NJ, and Princeton University supplement the research efforts of the big pharmaceuticals to create exciting biotech companies. There are 60,000 people working in the life sciences industry in New Jersey, which is the second highest in the country. We are excited to be part of that.

Mr. James T. Gunton has served the technology industry for nearly twenty years, initially in Silicon Valley as a rapidly promoted manager at Oracle Corporation, then subsequently in an investment capacity. In 1994, he joined Edison Venture Fund, and as vice president and principal he led its investment activity in New Jersey.

Mr. Gunton is active locally and nationally as a recipient of New Jersey's "Financier of the Year" and "Top 40 Under 40" recognition, and as a board governor of NASBIC, the private equity's trade industry association. He received a B.S. from Stanford University and an M.B.A. with distinction from the Fuqua School of Business.

Finding Success in Biotech

Johnny Humphreys
Board Chairman and Chief Executive Officer
Gen Prime

Creating Successful Deals

We create successful deals when we determine the needs and motives of companies with capabilities that complement ours. When we approach a deal, our goal is to extend our breadth and reach, which might mean acting as an arm of a larger company or having them extend our reach. As the saying goes, "One man's ceiling is another man's floor."

To locate potential opportunities, we always monitor the macro forces in our served industry, as well as in our own industry. We always seek opportunities to get to know all of the companies in our space. We obviously use the Internet as a tool to get up to speed on any given industry or company. We also read annual reports and interview customers for descriptions of the companies. This helps us to learn about the companies we might enter into a deal with.

In approaching negotiations, we contact companies that offer the best fit, and then we exchange ideas. During this time, we openly discuss the possibility of a long-term win-win situation. We ask the company's negotiators to explain their objectives, and we always ask more in-depth follow-up questions. This helps us determine their motivations.

In entering deals, we want results on our investment. If our desired result is customer access, we set definite targets and timeframes, and we monitor effort. We measure and learn what works, even if it differs from our original idea.

We decide a deal is worth the risk if the risk/reward ratio is favorable. The more a company ventures into the complete unknown, away from its core competencies, the more risk it takes, and the reward must compensate. If you can't afford lost opportunity time, follow a rule of golf, hockey, or basketball: "take the shot you know you can make."

We walk away from a deal when we realize the objectives of both parties do not align anymore. Frequent mistakes occur if parties only consider short-term objectives, or engage in win-lose negotiations. The deal can't be treated as a walk-away sale; both teams must be involved. You can create a good track record in deal-making if you work alongside good deal-makers and use vision, always remembering it's easier if your partner already shares that vision.

The best advice I have ever received about deals is that many of them do not stand the test of time. Often, the advice I find myself sharing with others is that you must build consensus within each organization as early as possible in the deal-making process.

As a "deal doer," I always remember that any major negotiation is a team negotiation. The leader must make clear the desired outcome and be sure everybody is helping to examine it from all sides.

My first rule for doing deals is to choose important objectives. The second rule is to choose the best potential partners. Third, you must sell to them as you would to a customer, but also listen to and understand their needs. You must determine if the goals are common for both parties, and if they are enduring. Finally, you must be creative in finding ways to meet the objectives of both parties.

Doing Biotech Deals: M&A and Partnerships

The best asset a biotech company can possess to convince another company to enter a partnership or mergers and acquisition (M&A) deal with it is a unique product suite or great momentum in the marketplace. Of course, possessing both of these assets is even better. The stage or size of the company may matter to the potential partner, but the two will match in some way. Culture, pace and speed, and public perception are three things that determine the best level of compatibility between two companies.

In negotiating with biotech companies, it's important to remember that, like software companies, they have nice margins and high expectations. Price, volume, and terms are all up for negotiation, and usually there is more than one combination of the three that can produce an acceptable outcome. Price and ego are two hot buttons that can kill a deal, and sometimes, large companies have in-house lawyers who will be so rigid they can make the deal not worth doing.

The chief executive officer usually takes the lead in structuring or approving the deal. The chief financial officer is sometimes creative, but usually more analytical. The chief technology officer is also analytical and must be on board with the deal, though he or she is usually not the originator. Outside investors, such as venture capitalists, and board members can make introductions or research potential partners after the deal's strategy is set. Lawyers and investment bankers usually provide "market" framework, and the best ones can consult on creative solutions.

The best way to do due diligence on a potential biotech partner or M&A candidate is to roll up your sleeves, talk to a lot of people, and listen without a filter. Possible red flags are past deals that didn't work, a lack of support in the organization, or other objectives that overwhelm the one under discussion.

Three valuation models are most often used for biotechnology companies. They are: discounted cash flow, multiples of cash flow, and multiples of revenues. For an excellent company, ten times the revenue is the usual acquisition cost. For a good company, six times the revenue (without premium) is the usual acquisition cost. The multiple paid times the net income is too big a variable to specify. It's determined specific to the company, growth rate, management depth, and market penetration. In determining purchase price, bigger companies have smaller multiples, because the growth rate of smaller companies is more exciting.

Biotechnology companies may possess other intangible assets, such as patents, which are notable in their absence. Some are very valuable, but many are pedestrian. A talented team of executives is a huge asset. In addition, momentum in a marketplace is very valuable.

When formalizing a deal, some aspects of the legal documents can come back and haunt a company. For example, the length of a contract could be a problem, as well as not having a provision for the resolution of practical difficulties. In particular, companies must be careful to avoid having assumptions about what the other party will do, without an explicit agreement. Not being thorough can lead to costly problems later on.

A biotech company can gain the most value from a partnership deal if it is treated like any other undertaking. It shouldn't be viewed as handing off an entire area of responsibility; it is managing the relationship with leveraged assets. In an M&A, it is important to ask how much more successful the acquired company can be with help from the acquiring company, not the other way around.

Four steps a biotech company can take to ensure an M&A deal has a positive financial effect on the company are: understand the bottleneck areas in your company, devise a way others can do a better job in some aspect in a bottleneck area, be sure the objectives align, and negotiate a deal that will work for both parties long term.

Four things a biotech company should look for in a partnership to ensure a positive affect are: a larger capacity for whatever you are supplementing, a culture you admire (somewhat like your own), a broad-based desire from both companies to make it succeed, and a win-win negotiation style.

Mr. Johnny Humphreys is the board chairman and chief executive officer of Gen Prime Corporation, a supplier of biological threat detection for Homeland Security applications. Humphreys joined the board in 2003 and became chief executive officer in March of 2004.

Prior to his current position, Mr. Humphreys was chairman of the board of Linesoft Corporation from January of 2001 until April of 2002, when it was acquired by the Itron Corporation. During this time, the supplier of design software tools for utilities doubled its revenues to $15 million. The company also raised $23 million in venture capital financing during that period.

Mr. Humphreys was chairman of the board of Itron from May of 1998 until January of 2000. He was the president and chief executive officer of Itron from 1987 until June of 1999. Under his leadership, Itron became a leading global supplier to utilities of systems *used for collecting, communicating, and analyzing electric, gas, and water usage. Today, Itron is the world's largest supplier of systems for the emerging automatic meter reading market, with approximately 600 million in revenues. Itronix is a spin-off of Itron, and the combined revenue of the two companies is approaching $1 billion.*

From 1975 to 1986, Mr. Humphreys held various executive positions with Datachecker, a subsidiary of National Semiconductor, including serving as president from 1980 to 1986. During that period, Datachecker grew from $12 million in revenues to $240 million per year. Datachecker was instrumental in creating the supermarket scanning business. Datachecker is now a part of Fujitsu/ICL, and the systems Datachecker invented are still a mainstay of the product offerings.

In 1986, Mr. Humphreys was appointed senior vice president of National Semiconductor's information systems group, and was responsible for strategic planning, programs, and partnerships for three operating divisions: National Advanced Systems, Microcomputer Products Group, and Datachecker.

In 1974, Mr. Humphreys worked in market development for EDS Corporation. In 1967, he founded Electronic Laboratories, Inc., now Telxon Corporation, a maker of handheld computers. He was president of the company until 1974, when Telxon merged into Symbol Technologies. He was manager, vice president of engineering, and executive vice president of Gulf Aerospace Corporation from 1963 to 1967, and worked for Texas Instruments and General Dynamics before joining Gulf.

Making Your Way Through Biotech Dealings

Roland H. Johnson
President, Founder, and Chief Executive Officer
Piedmont Pharmaceuticals LLC

C-LEVEL BUSINESS INTELLIGENCE · C-LEVEL BUSINESS INTELLIGENCE
ASPATORE BOOKS

The number-one golden rule for participating in a biotechnology deal, a merger, and/or an acquisition is to fully understand where you are in the process of building your company (i.e., where your company is in its development stage). Is it early stage? Is it late stage? Once you have decided to do a deal, you have to be prepared to fully commit your company's energy and resources to completing the deal. The entire team has to be involved in preparing for the due diligence, including getting all of the information that might be needed. You must be forthright on everything. Make sure everything is on the table, even things you may not think are important. Be certain everyone on both sides of the deal has full disclosure about your technology. Make their due diligence easy; or, if your company is the licensing or acquiring company, prepare your team for the process in order to make your due diligence easy and efficient. During the course of the deal, look for the gaps. Look for things that are inconsistent with what you are being told, things that change from one meeting to the next, or pieces of information that haven't come out in the process. Do this at every step, from the initial meeting to the moment you sign the paperwork. If you are doing a deal, particularly a licensing deal or an acquisition, make sure you're getting what you think you're getting. Many deals establish future success measurement benchmarks that tie the other company's success with your technology, or your compensation with their future success. Ensure that you are dealing with people you can trust, and build incentives into the process that reward cooperative trust. The process of evaluating companies for potential deals is analogous to going on a date. However, reaching and implementing a deal and making it work is like marriage. It takes a longer-term commitment, and you want to make sure you laid the foundation to make it a good one.

When to Walk Away

Misunderstandings happen. Everyone has different motivations and different past experiences that influence behaviors at the bargaining table. If you begin to find inconsistencies, you need to address them by raising your concerns, and then evaluate how the other party responds. If the response isn't appropriate, go up the food chain. Make sure you are talking to the people who have the power to make the decisions. If there are misunderstandings, make sure you truly understand what the issues are, and work to develop processes that will minimize future misunderstandings. Try to look at the issue or concern from the perspective of the other company. Put yourself in their shoes. You must strive to reach a reasonable negotiating position that attempts to address each company's needs in a way that results in mutually aligned incentives that reinforce each company's motives to succeed. Families, husbands, wives, and children have miscommunications — it happens. It's no different in business communications. When you are working with a deal, miscommunications can and will take place. Strive to make sure that when issues come up that may be misinterpreted between the parties, you attempt to do all that is reasonably possible to ensure that communications are clear, concise, forthright, and timely.

When do you walk away? If you confirm that the other company is not being forthright, particularly after you've qualified your questions, re-qualified your questions, and moved up the chain of command, you should consider walking away from the proposed deal. Unquestionably, if you find out the other party is intentionally misleading you, walk away. I have done it, and I would do it again. Walk away even if you desperately need the deal, because in the long run, staying in the deal is going to come back to hurt you.

Determining People's Motivations

First, do your homework. You can find a tremendous amount of information on publicly traded companies online. Identify and study any public information you can find on the other company and its management. Evaluate the other company's areas of research. What is it trying to achieve? Look at its successes and failures. Has it had problems with other partners? Is this a consistent pattern? What is the state of its financial health? Try to learn what its past and future prospects are, and about its management team, with a particular focus on the company's strategic and tactical strengths and weaknesses. Identify and talk with outside consultants or investors that may have insight on the company and its technology. These are important factors to consider and evaluate before moving to face-to-face meetings.

Meeting with people takes time, but at the end of the day, face-to-face discussions will provide you with your most important insight on the other company. Throughout the course of meetings and negotiations, you should maintain or expand your internal and external efforts to gain insight on the potential partner. If you successfully close the deal, there is a good chance you may be required to work with many of the same people you are now negotiating with for years to come. Remember, you're entering a business marriage. Because of this, you need to do everything possible to confirm that your potential future partners have a passion to win.

Spotting Opportunities

You need to establish clear strategic objectives, assess your core competencies and strengths — such as technology, people, manufacturing, marketing, sales, or capital — and effectively identify your strategic and tactical needs. By doing this, you can effectively design a first-cut internal screening process for identifying opportunities. Your desire to enter into a deal should be driven by your company's strategic objectives. Deal objectives cover a wide range of needs, from technology to capital. You may hear people make comments that refer to a "typical technology licensing deal," however, my experience leads me to conclude that all technology deals are different and that the only thing typical is that they involve technology. The technology, the people, the capital requirements, the development stage of the company, the intellectual property, the time to market, and the market potential of every deal is different. If you are in- or out-licensing technology, you have to know your needs. Some of the most straightforward deals involve combining two or more different technologies to develop a more effective product than either of the technologies alone, for example, pharmaceuticals and delivery technologies or medical devices and pharmaceuticals.

In the end, the potential for successfully capitalizing on an opportunity is almost always linked to the party's ability to effectively align the incentives of the deal. The true success of a deal is directly linked to the company's ability to create value for the stakeholders, and as a result, effectively capitalize on the market opportunities that stem from the transaction. Successful execution of the value creation process is often directly linked to how effectively the companies were in combining their mutually aligned incentives during the deal process.

M&As and Partnerships

My background includes in-licensing technology, out-licensing technology, investing in established companies, and selling companies in life sciences. The consistent theme of every deal I have ever participated in is that value creation is always ultimately driven by needs in the marketplace. What does the market need? Are you really solving a problem that needs a solution? Technology or products that do not ultimately address the needs of society will not build significant sustainable value. You should look at technology from the perspective of the marketplace and ask, "What does the marketplace need? What is the size of the opportunity?" Then determine if your idea, technology, or team can address that need, and if the size of the opportunity warrants your entry.

Following are three keys to maximizing your success:

1. One of the most straightforward ways to identify a market opportunity is to look at established, fragmented markets that do not offer a satisfactory remedy. As an example, let's look at a product concept for consideration. Let's use the allergic reaction that occurs to poison ivy as an example, something people rarely think about unless they have it. If someone develops blisters from poison ivy, they go to the drugstore where they will find many different products, none of which will get rid of their symptoms in two minutes or even two days. They will likely continue to have severe reactions to the poison ivy for three to five days, no matter what medication they use. People would love a treatment that would take care of the problem immediately with no side effects and that is pleasant to use. I like to look for markets that offer several different products with no clear market leader. Often, the mere fact that there are multiple products tells me they do not effectively address the customers' needs. The available products are better than nothing, but they don't offer a true solution. If you could take the poison ivy treatment market with all its fragmented products and come up with a solution that would give you immediate relief and immediately eliminate the itching and lesions, you would have a winner.

This type of theoretical product would allow you to consolidate the market and change the basis of competition in the poison ivy market segment. If you can consolidate twenty different products, all of which are doing X number of dollars, you may have a significant product on your hands, one that might justify significant investment and allow you to change the value proposition in the category. The poison ivy example is similar to the approach we are taking at Piedmont Pharmaceuticals to address the market for head lice treatments. We are developing a treatment that is expected to be the first FDA-approved drug for the treatment of head lice that does not use a classical insecticide as its active ingredient.

2. Strive to develop products that have high barriers to entry. You want patents. You want a registration process. Once you have developed a market, you don't want another company to come in and displace your product by riding on your ideas or by using the same or a similar technology. You would prefer to have a product that, once you have made the investment and commitment, is going to allow your company to take the leadership position in the category.

I used treatments for poison ivy and head lice as examples. What if someone had heart disease, for example, and could take one pill that would cure heart disease instead of going through years of treatment and surgery? You can dream about a possible treatment of this type and easily conclude that the idea is crazy, and could never happen. Yet, over the years, a lot of products have been developed that people assumed could never be developed and that are now used every day to address customer needs. Just because it hasn't been done, doesn't mean it can't be done. The market for this type of product would be huge. A pill that could cure heart disease would be an industry shifting type of product. It would affect the entire medical industry. If a company had this type of breakthrough and good intellectual property, it would be in a position to build tremendous value with limited amounts of proof of concept research. This is a company that should be in demand, from partners that want to joint venture, license, invest, or acquire the technology or the company.

3. Build a strong team. The strength of a company's team can make it more or less attractive as a potential partner. Many companies, particularly large

ones, have difficulty swiftly moving life science products from research through the development process to commercialization. If you have a team of people with proven track records of developing breakthrough products that address market needs, you would have a valuable resource that others may want to acquire or access.

Other Factors (Size, Stage, and Benefit to Society)

The size and stage of the company matter greatly. If you are talking about a merger or acquisition, for example, value creation is driven by profitable positive cash flow. Companies that are looking for acquisitions seek to secure assets with accretive cash flow. Companies want to acquire profitable, cash flow positive companies, with great teams, that offer new intellectual property with the potential to change the base of the competition within a targeted strategic segment. If your company has this group of attributes, you should be in a position to negotiate the terms of an acquisition as a buyer or seller. The value or strength of your company can be enhanced further, subject to the development phase of your new technology; from early research screening phase to product launch. (The example above refers to a type of M&A deal that may offer the potential for value creation. However, there are examples of deals that lead to value creation that are based primarily on intellectual property and early-stage research. An example of an early research deal could be based on the theoretical heart treatment pill that was presented above.)

I feel the value of a deal for the overall benefit of society is a factor to consider in addressing the company's objectives as they relate to a deal. I personally feel that whenever a company is considering two equivalent strategic investment programs with similar risk and benefit profiles, it should ultimately select the investment that has the greatest potential for making a meaningful improvement in the quality of people's lives. Quality and life fulfillment issues are difficult to measure; however, these factors are important and should be included in the deal assessment process.

Rules for Negotiations

All parties should diligently strive to assure that all aspects of the deal process are based on honesty and integrity. Every effort should be made to ensure that when information is exchanged, the receiving party can easily differentiate between information that is based on scientific evidence versus opinion or hypothesis.

As you begin to consider a licensing/acquisition/merger opportunity, you should review or reassess and reconfirm your company's business objectives from both a strategic and tactical perspective. You should ensure that the technology under consideration for licensing or acquisition complements your overall objectives and does not potentially undermine the strength of your company. No one wants to lose access to a technology that might prove to be critical to the company's growth. We need to once again consider the stage of your company's development in light of its strategic objectives. What does your company want to be when it grows up? For example, do you intend to build a major pharmaceutical research, development, marketing, and sales company, or a contract research company? Do your objectives allow for a strategy that targets its initial growth as a contract research company, but with a longer-term objective of becoming a major pharmaceutical company? Is the technology being considered for out-licensing or in-licensing critical, or is it an incremental complement to your strategic objectives? These questions should be considered in light of the depth, quality, and maturity of your company's overall technology portfolio, and the company's current and future capital requirements.

Once you have confirmed that the benefits of licensing, acquiring, or selling the technology outweigh the risk, you should aggressively move the licensing/acquisition process forward. Quality due diligence is critical to the success of the deal for both parties in order to minimize risk and ensure success in creating value for both parties. If you are considering a technology or transaction structure in which you have limited experience, you must find and engage the best technical people possible to adequately assess the opportunity. It is also important that you engage the best legal, financial, and accounting advice possible. Every decision has risk: you must attempt to minimize as many of the risks as possible in the due diligence process. You must dig deep to identify potential oversights and/or discrepancies that may undermine the proposed acquisition or licensing opportunity. In addition, you should work to identify areas for incremental value creation that may not have been recognized and/or may add value to your existing products or technology. It's not difficult to find examples where a company has done an excellent job of conducting and developing its technology, but failed to appreciate the difficulties and complexities that later undermined success, perhaps due to manufacturing issues, conflicts with intellectual property, or a wide range of other issues.

We have addressed many of the issues that should be dealt with in the negotiating process. I would now like to focus on the issue of negotiating the structure of payment for access to the technology. Again, consideration must be given to the development stage and financial strength of your company and your internal assessment of the value that may be created with your technology. (Do you believe in the technology? What is the cumulative long-term value of the technology?) There is a wide range of creative approaches that can be used to address compensation compared to risk and future value creation. An early-stage company with limited access to capital may tend to focus on short-term financial compensation at the expense of long-term compensation, especially if the technology is not strategically important, or the company does not fully understand the long-term value (market potential) that may stem from the technology. Upfront payments of cash may be discounted significantly from the future value that may be created with the technology. The entire process of negotiating value is a process of tradeoffs of short- and long-term needs versus risk and opportunity. The seller or licensee should strive to reduce the risk associated with the deal for the acquirer, in order to secure the maximum return from a transaction. There is a direct relationship between the effectiveness of reducing risk and the level of compensation received. If you believe in your product or technology and in your potential partner, you should strive to benefit from their upside success. Given the right partner (with sales, marketing, distribution, research and development, financial, or other strengths), you may desire to structure compensation in a way that allows for maximum upfront payments (based on financial models) and participation in future accretive value that should flow to the licensing or acquiring company by way of equity, performance incentives, royalties, and/or transactional liquidity events.

The CEO's Motivation

The chief executive officer's (CEO) motivations and objectives should be driven by the tactical and strategic objectives of the company's stakeholders (investors, boards, employees), the developmental stage of the company, and the type of deal. The number-one job of all CEOs should be to build value for the shareholders. The CEO's approach and aggressiveness in driving growth and taking risks are often linked to the investment profile of the company's board of directors, board of advisors, and investors (friends and family, angels, venture capitalists, public markets), which in turn should be tied to the company's strategic business objectives. The approaches used by the CEO to build value may vary widely based on the type and developmental stage of the company. Is the company privately held or public? Is the CEO the sole investor, a significant investor with a limited number of partners (friends, angels, or venture capitalists), or a minor investor with many private or public investors? Are the investors short- or long-term investors?

The CEO's job is to lead the team of employees and investors. The CEO is responsible for establishing, communicating, and securing support for the company's objectives. The CEO ultimately determines the importance of moving a proposed transaction forward while trying to minimize risk through the input of advisors (technical, legal, financial, regulatory, employees, and the board). The CEO should look at everything the company does with the knowledge that the investors have a clear understanding of the company's business objectives and that the CEO has a clear understanding of the investors' objectives with regard to short- and long-term growth and/or exit objectives. CEOs are motivated by creating the greatest financial rewards versus risk for his or her investors, and ultimately for the CEO. The CEO's performance objectives and financial incentives should be mutually aligned with the investor's financial objectives in order to build a framework that drives maximal value and minimal risk for the company. The CEO should drive the vision and performance process forward by aligning all employees' objectives and incentives with the company's objectives.

The CFO's Motivation

The chief financial officer's (CFO) role is second only to the CEO in his or her responsibility to protect the investors' capital and creating value for the company. The CFO and CEO have clear fiduciary responsibilities to the investors in assuring that the company's accounting procedures are in compliance. The CFO is responsible for assessing all financial aspects of a potential deal transaction, including but not limited to financing, deal structure (equity, stock, royalties, milestone payments, etc.), and taxes. The CFO should be a key member of the company's deal team. The CFO's responsibility must always be consistently focused on the investors and the company's overall value retention and creation process.

The Roles and Motivations of Outside Investors

Investors are driven by their desire to secure the highest return on their capital versus risk, as compared to other investment options. All investors (angels, venture capitalists, private, public) have a right and responsibility to ask questions and secure prompt and straightforward responses. The company should drive to establish processes for providing strategic business and financial updates to the investors on a regular basis. Communications are always warranted with a significant change in the company's strategic objectives. In addition, investors should be notified of significant changes in the company's business outlook (positive or negative) that may materially impact their investment. Most importantly, deals will have a noteworthy impact on investors, and as a result, they should certainly involve the support of the board of directors or advisors. A pending deal should be communicated to the investors prior to completing the transaction, unless the deal might be compromised due to confidentially concerns. My experience shows that informed investors are supportive investors (identify new investors, provide strategic insight, open doors, and provide supportive business contacts).

Venture capital investors can play a critically important role in initiating and completing deal transactions. Professional venture capitalists should bring significant value (above and beyond capital) to a deal. Major venture capitalists have professional, proven deal teams in place that can guide or lead the deal process. In addition, the venture capitalists should add a level of overall credibility to the deal process as it relates to the company, its technology, its team, and its financial stability that is the equivalent of "The Good House Keeping Seal of Approval." The involvement of a venture capitalist may further reduce the perception of risk in the deal process, and as a result, might drive a higher transaction value from a deal.

Lawyers and Investment Bankers

Lawyers should strive to protect the company and investors by insuring that legal agreements (term sheets, contacts, etc.) provide the enforceable terms and conditions that will be used to assure compliance with the terms and conditions of the deal. Term sheets and contracts should provide legal clarity and minimize ambiguity with regard to the current and future responsibilities, implementation, and enforcement of the agreement. Agreements should reinforce clear confirmative communications between the parties with regards to commitments and obligations under the terms of the agreement. Investment bankers should be called upon to identify, develop, and implement the financial structures that will support the company's efforts to complete the deal and allow the company to meet its business objectives. The investment banker's role is to insure that the company has access to adequate capital and the type of debt structure that may be required to complete and support the proposed transaction.

Due Diligence on a Potential Biotechnology Partnership or Merger

The companies should reach agreement as spelled out in a term sheet for a proposed transaction before moving from informal due diligence (review of public information) to formal due diligence. Both companies should be prepared to enter into confidentially agreements as discussions move forward. The parties must feel comfortable sharing confidential information with each other in order to move the due diligence process forward as quickly as possible. Each party should attempt to minimize the disruptive effect of due diligence on ongoing business operations to the extent possible. The parties can most effectively address this objective by providing clear, forthright, unencumbered access to information. Due diligence is similar to a scientific experiment, in that if you conduct the experiment right, someone else should be able to replicate it easily. In due diligence, you try to go through the other party's thinking process in an attempt to confirm the other party's conclusions. Effective due diligence will require your company to have access to a wide range of different talents; intellectual property, regulatory, manufacturing, finance, marketing, sales, and many others. Your company should develop an outline of the key issues or questions that need to be addressed in the process, and identify the people that will be required to adequately assess in a timely and efficient manner. How much capital will be needed to complete the transaction and commercialize the technology? How many years will it take to bring the product to market? How quickly can your company complete an in-depth competitive analysis of the technology today and other technology or ongoing research in the field? What is the potential for the product? When you bring this new product to market, you don't want to find out that four other companies are bringing the same or a similar product at the same time. You must be certain your company has diligently managed its assessment of all of these issues as effectively and quickly as possible.

Valuation Models

The valuation models for biotechnology companies are all over the board and really depend on the market potential of the technology, the proprietary nature of the deal, the length and quality of the patents, and the financial models. Instead of simply looking for a rule of thumb, I would find the best financial analysts possible to help identify and confirm the value of the proposed transaction.

Common Legal Pitfalls

Every part of a legal document can come back to haunt you. With that said, my advice would be to get the best legal talent possible. The value proposition that can stem from a simple technology licensing deal can change with time. A simple deal four years ago might be tremendously valuable today. A small, relatively simple, quick licensing deal can come back to haunt your company. In order to save on resources, companies will enter into these types of "straightforward" agreements with local counsel without review or input from intellectual property counsel. Quality local counsel is critical to your company's ongoing day-to-day legal work. However, they may not be capable of providing the level of finesse that may be necessary for intellectual property and regulatory issues that support technology-based life science products.

Steps to Ensuring the Best Financial Impact

High-quality intellectual property and high-quality intellectual property assessments are essential. Do not skimp on the quality of the work or the quality of the law firm you use to develop and protect your intellectual property. Intellectual property is one of the most valuable resources an early-stage biotechnology-based life science company can own. Many early-stage companies don't have significant revenue streams from ongoing operations, and as a result, they do not have manufacturing facilities, sales forces, or established brand names to protect the company and build value for their investors. At the end of the day, these companies rely on their intellectual property patents to build and sustain value for their investors.

Capital is always a critical concern for early-stage life science companies. There is seldom enough capital available to move research and development forward in a unified and expedited manner. Companies always need additional capital, however, they don't always need it all up front. A lot of people make the mistake of thinking that unless they have X amount of cash in the bank, they can't accomplish anything. If your company takes this approach, it may never get anything done. Early-stage companies will never have enough cash to feel comfortable. General George McClellan, commander of the Union Forces during the Civil War, provides a good example of what not to do. McClellan's forces and supplies greatly outnumbered General Lee's Confederate troops. However, General McClellan never thought he had enough troops and supplies to move forward aggressively, and as a result, he was relieved of command by President Lincoln in 1862. Your company must commit to always moving its value creation process forward: the only way to move that value creation point forward is to at least take incremental steps in research. Your company must move forward, even if the pace isn't as fast as you would like. Every successful study should add value and credibility for the company. If that means you can only afford one study and it proves your hypothesis, you've strengthened your intellectual property base and improved the viability of your company. One successful study may improve your ability to negotiate and secure the support of venture capital or angel investors. The study may strengthen your ability to talk to potential licensing partners. Your company may have very talented researchers, good ideas, and solid intellectual property. However, if the development process is not moving forward, the company may prove to be of limited or no value in the future. Your company must move forward and demonstrate that it can continuously build value.

You must build a dedicated team that will not quit. In the early stages of a new product or technology, when you don't have adequate capital, quitting is often easier than continuing. You must build an atmosphere that allows the team to see that if the company or product wins, they will win. The team should be built and supported by an incentive system that provides equity and instills ownership in the company, the process, and the products. Each member of the team must be able to clearly see how they can and will participate in those long-term rewards. Each member of the team is taking a risk in taking a job with limited short-term security versus the perceived security that may be offered by established companies or research institutes. The members of the team need to know that if the risk pays off, they will have the potential for greater financial rewards than they might have had access to in an established organization.

In summary, I feel that creating and protecting intellectual property, pushing the value creation process, and building passionate committed team members is the key to ensuring the success of early-stage biotechnology and life science companies. By effectively addressing these issues, your company should significantly increase its attractiveness to sources of capital, which in turn will hopefully lead to great products that meet the needs of the market and society.

Mr. Roland H. Johnson is the president, chief executive officer, and founder of Piedmont Pharmaceuticals, LLC. Piedmont develops novel pharmaceutical solutions to address patients' medical needs. Piedmont focuses on addressing human medical problems by utilizing its patented platform drug delivery technology. In addition, Piedmont is a leader in developing parasitology-based pharmaceuticals. Mr. Johnson is also the chief executive officer and founder of Triad Specialty Products, LLC. Triad is a wholly owned limited liability company of Piedmont Pharmaceuticals, LLC with a focus on developing unique veterinary products.

Previously, Mr. Johnson was the vice president and corporate officer for IDEXX Laboratories. IDEXX purchased Blue Ridge Pharmaceuticals in 1998. Roland continued as president of the veterinary pharmaceutical subsidiary of IDEXX Laboratories until July of 2001. As president, he was responsible for building and establishing IDEXX Veterinary Pharmaceuticals. In addition, he served as a corporate officer and vice president of IDEXX Laboratories. Prior to this, Mr. Johnson served as the president, chief executive officer, and founder of Blue Ridge Pharmaceuticals, and as the vice president of Ciba Geigy Animal Health (Novartis).

Mr. Johnson earned his M.B.A. from Central Michigan University in 1983 and his B.S. from Western Carolina University.

Keys to Success in Biotechnology

Thomas G. Klopack

Consultant

C-LEVEL BUSINESS INTELLIGENCE · C-LEVEL BUSINESS INTELLIGENCE · ASPATORE BOOKS

Personal Philosophy

Business deals in biotechnology are usually an important element in a company's strategy. A deal with a large pharmaceutical company or another biotech can bring in significant non-dilutive funding, serve to validate the technology to other potential customers, and validate the business model to investors. In 2004, according to Recombinant Capital, over $30 billion worth of biotech deals were signed. This is a threefold increase over the number of deals done in 1994. Deals can range from early-stage discovery science to actual drug molecules. From 1995 through 2004, average deal terms, including over 19,000 deals, broke down to $25 million for lead molecules, $40 million for preclinical compounds, $60 million for Phase I compounds, $60 million for Phase II compounds, and $110 million for Phase III compounds. Each deal can and usually does contain several components: upfront payments averaging $10 million, research and development funding averaging $35 million, milestones averaging $100 million, and equity averaging $20 million. Not all deals contain each of these elements, but most contain at least one of them. It is also interesting to note that the median deal terms were less than half of the averages. So, most deals are significantly smaller than these averages, and a small percentage of deals are much larger.

Some facts about the biotechnology industry are useful to set the context in which deals take place. There are roughly 3,000 biotechnology firms, but there are only about fifty companies that have significant resources — cash flow — to do major deals. A major deal in the industry is at least $500,000, and generally tens of millions of dollars. Merger and acquisition (M&A) transactions and major drug product licenses can run into the hundreds of millions, as outlined above. The companies that have the appetite and resources for that kind of a deal are the big pharmaceutical players. In recent years, it should be noted, the number of players in that category has gotten smaller as there have been consolidations. Some biotechnology firms have graduated into the top category, such as Amgen, Biogen, and Genentech, but some of the major players — notably Parke Davis and Pharmacia — have been acquired. This has reduced the number of significant companies with the resources for major deals. In 2004, there were approximately 400 deals signed, with some companies signing more than one. That's not a large number considering the 3,000 biotechs that are out there. Closing deals distinguishes biotechs in the industry.

Consequently, the biotech industry deal environment is very much a buyers' market. All the large pharmaceutical and biotech companies are trying to figure out what their internal requirements are, and what products and technologies they want to acquire to meet those requirements. They have business development personnel that collect internal requirements, source products, and technology, and set the structure for deals. These business development individuals are guided by the broad science and operating community of the company, which tells them what requirements fit the strategy the company is following today and help it move forward into the future.

There is clearly an appetite for licensing successful drugs. Large pharmaceutical companies have the marketing and sales organizations to effectively get drugs into the hands of doctors. The major problem faced by both pharmaceutical and biotech companies is that they are not developing drugs and getting drugs approved fast enough to meet their growth rate requirements. This issue is well documented; even with increased spending on internal research and development, the pharmaceutical industry has not been able to increase the success rate of finding drugs. So they are increasingly interested in new drugs developed by biotech, and they are increasingly interested in these drugs at earlier stages of development.

Most biotechnology firms, however, are not selling drugs — they are selling technology. And historically they have sold technology related to early drug research called "discovery," rather than the later stages of clinical "development." Companies selling technology have to have a compelling technology that adds value to the buying company, usually by helping to discover and develop safe drugs faster. If it is not a drug, it has to be a tool that the pharmaceutical company cannot live without, that would help them dramatically, or that they cannot allow their competition to have. These types of technologies have been few and far between, since the drug discovery/development process is long and highly variable, which makes proving the value of a technology difficult without actually finding a drug.

Even with a compelling technology, the biotech company has to be a real company, not just a science project. Biotech companies can be, and many have been, focused on solving interesting problems, not on being real businesses. Real businesses develop and make what the market wants, sell it at a profit, and deliver it on time and working. Pharmaceutical companies will hesitate to do a major deal involving major dollar values unless the biotech looks like a company with an organized team of competent people, and the money and resources to deliver on the deal. These are deals involving multiple millions of dollars and usually multiple years. A biotech company must be able to show, during both the due diligence and execution phases, that it can actually carry out the contract.

Deal Process

Success is initially defined as closing the deal, usually punctuated with a proud announcement. However, large deals usually involve timeframes and performance targets tied to milestone payments, whether it is finishing development and successfully transferring a tool, or successfully completing drug development phased trials. Both initially closing a deal and successfully executing it usually have a specific process. These processes are generally well understood in more mature industries, but have been sometimes not well acknowledged in biotech.

The process to sign a deal usually starts with contacts between a biotech and a set of target pharmaceutical and biotech partners, most often initiated by the selling biotech. These early discussions raise awareness in the customer company and try to make the connection of the biotech product (technology, tool, drug, etc.) with the customer pharmaceutical and biotech requirements. Multiple discussions might be assigned a slight probability of success, say 5 or 10 percent, since multiple discussions indicate some interest. This is in the context of those fifty companies being continuously bombarded by 3,000 biotechs trying to do deals. As the discussions progress, a term sheet is usually prepared. This outlines what will be provided, the general financial terms, time for execution of the deal, general ownership of intellectual property, and other key matters that both supplying biotech and customer pharmaceutical/biotech want to have clear and generally agreed to up front. Until there is a term sheet that has been agreed upon, a deal is usually not close to being done. At the term sheet stage, there is a 30 to 50 percent probability, at most, that a deal will take place. It is to the biotech's advantage to get to this stage and get their demands on the table and discussed, if not actually finally agreed. Term sheets are usually not binding documents, but they define the direction the contract negotiations will take. If the terms cannot be agreed upon, getting a contract in place will be very difficult.

The next step in the process is a draft contract. The drafter of the contract has the opportunity to set the structure and tone of the negotiations. Once a draft is in place, based on the term sheet, and has been recycled by both buying and selling companies, the probability of a deal happening rises up to 75 percent. This requires that a contract is very close to final form—all key terms and key language are agreed upon, and the buying company has done several reviews at high decision-making levels. This includes legal reviews, management reviews, technical and patent due diligence, multiple meetings and visits to the selling biotech, and often a preliminary smaller exploratory deal. No major deal is done without significant legal, intellectual property, and scientific due diligence, so expect it and push to get it done up front.

A deal is never done until it is done—that means it has to be signed and approved by both parties. Even when a contract has been given a verbal go-ahead and there are all indications that everything is moving forward, it is still not done until it is signed. Depending on the company, the contract may have to go to the board, or at least beyond the division or site with which the seller biotech is talking, and this can mean another layer of examination, and more questions and concerns. Until it is officially signed, there is always something that can happen to make a deal fall through. The buyer can be caught up in acquisition talks of their own, an FDA action can take place, the investment market can take a turn, or a key manager leading the process for the buyer can leave. Imagine the effect September 11, 2001 had on deals in process.

Once a deal is signed, the process then moves to execution. Most large deals involve significant performance milestones and/or royalties that the selling biotech is required to perform in order to get total value from the deal. There are "real dollars" associated with deals and "biotech dollars." Real dollars are the irrevocable upfront payments made at deal signing. Biotech dollars are the payments requiring performance under the contract. Biotech dollars are usually several times, if not an order of magnitude, more than the real upfront dollar payment at deal signing. See the statistics quoted earlier from Recombinant Capital. Executing on the contract is the key to getting its full potential, and usually key to a biotech's success.

Executing the deal requires a real company with a competent team and the resources to carry out the deal. It will also require the technology to work correctly, the drug to continue to be successful in human clinical trials, completed development and potentially key improvements to a tool, successful transfer of technology to the buying company, or the ability to scale up and produce the product. It also usually requires the buying company to continue to be interested in the deal, something that can change with changes in market conditions, the intellectual property landscape, or even the buying company remaining independent. Some of these factors are totally in the control of the supplying biotech. Some factors are totally outside the biotech's control. All factors can be considered in the contract stage and appropriately dealt with given foresight of the potential situations that can arise. Execution depends both on doing a good job after the contract is signed, and anticipating and providing methods of dealing with all the potential serious future scenarios in the contract itself.

Planning Deals

Planning for deals starts with a realistic business plan that takes into account the importance of deals in the biotech company strategy. Deals may be an important source of funding, of validation of the value of the technology to investors or other buyers, or as the revenue stream that leads to profitability. Each of these reasons can change the nature of the deals that are prosecuted.

If the business model plans to get significant revenue by selling to large pharmaceutical companies, the biotech must develop what multiple pharmaceuticals want, and it must structure deals that allow free sales to multiple customers. If the objective is to develop a product and sell it to a customer pharmaceutical for validation of the technology, then the deal may be a rare one-time event, and it might be done as an exclusive deal. This deal would be done to gain investor confidence and funding, which could then be used for developing novel drugs for a later, higher value deal or building wholly owned sales.

Many biotechs have developed interesting science, but not necessarily products or technologies that pharmaceuticals want to buy. The drug discovery and development process is well documented, and most buying pharmaceutical companies try to communicate their interest in aspects of the process to the biotech community. If a biotech can develop products that solve these outlined problems and prove that they do, they will usually get interest and business from pharmaceutical buyers. For example, a key problem in drug development is the fact that development trials are very expensive and time-consuming, and human beings are complicated creatures that have not been successfully modeled for trials. If a biotech has a technology that raises the probability of drugs getting through development, raises the probability of early failures (so losers can be eliminated), lowers the amount of time needed, or lowers the number of people needed to test in any trial, that technology/product directly solves a well understood problem and is worth significant value.

Any biotechnology company must understand what value they are adding to customer companies by understanding what their industry needs, and how their product/technology addresses the needs in comparison to other methods. This is usually easy to do with a new drug acting on a new target in a major disease market, but it is harder to define the further away from the final drug the problem is that the product addresses. So, late-stage drugs that have successfully completed Phase II trials are worth more than early-stage drugs at Phase I. Whether the product is a tool, a technology, or a drug, it is always key to be able to make a strong, economic argument that the proposed deal makes eminent sense to the customer. More so today than ten years ago, biotechs have to show significant value for the money that will be spent by the pharmaceutical/biotech customer.

Negotiations

The first goal in doing a deal is to get a buyer's or multiple buyers' interest. Once that is done, the biotech is on a process to formulate a term sheet and a contract. Biotech deals are usually large individual events, and as such are customized to fit the customer's needs, wants, and biases. From a negotiating standpoint, identifying multiple buyers benefits the selling biotech immensely. The aspect of a bidding war usually benefits the seller, particularly if the deal involves exclusive rights to a technology, drug, or the entire company. This can be done even when negotiations have started with one company.

The biotech should start off with strong economic arguments that fully value the technology/product offered. Most biotechnology companies do deals on critical expensive technology too cheaply — like announcing, "will discover drugs for food." The selling biotech must meet the objectives of the deal, which almost always involve making a profit on the deal. Successful deals are win-win situations: the buyer must really want what the biotech offers, and pay adequate value for it. This starts by setting the expectations of value at the term sheet stage. Asking for too much is usually better than asking for too little at this stage of the negotiation. The deal must be a good event for the biotech — profitable, carrying the business model and strategy forward. There are some exceptions: for instance, if the biotech is trying to make an initial deal to show credibility for a technology, to obtain a user, and publish on the user experience, or to gain critical financing to bridge to a better financing climate or later significant company milestone.

Biotechnology and pharmaceutical company deals have been unique among many industries in terms of how much technology has been sold for low prices — well below development costs or even value to the buyer. Biotechnology firms in general have been too desperate to get deals, and have not initially asked for and held out for the real value of the technology or drug. One underlying cause of this is a faulty business plan — one that does not build a viable business. If the business plan is flawed — not calculated to build a sustainable company — then deals based on it will be flawed also.

One important strategy during negotiations is to deal with the highest level within the buying company as soon as possible — get to the decision-maker quickly. Pharmaceutical companies are large and can be layered with multiple groups evaluating deals but having no real decision-making authority. Getting the decision-maker involved, understanding the process the buyer has for making a decision, understanding the budget available, and timing for it will all both push forward a deal and make the real drive and timing for the deal clear. For multi-million-dollar deals, you certainly want to deal with a key executive, the vice president level, or members of the managing team of the buying company. If you are not at that level, you are not talking to a decision-maker. If a technology or product causes change across multiple organizations in a company, the decision authority will necessarily be higher and cross-functional. Aurora Biosciences was selling a system that cut across multiple departments to dramatically impact high throughput screening. This required the sale to be made to a high-level decision-maker in the organization who could reallocate resources among the affected departments. The large drug discovery/development deal between Vertex and Novartis in 2000 required the chief executive officer of Novartis to outsource spending of research and development funds to Vertex, rather than scale up internal research and development. This kind of decision would not have been made at a lower level.

The negotiating team for the selling biotech is important. The lead business development person will usually do the initial contacts and manage the process. But since doing a major deal between a selling biotech and a buying pharmaceutical can have such a large impact on the biotech, the entire management team will usually be involved to some extent. The buying pharmaceutical will also want to see the involvement and commitment from the management team. It is another indication of the willingness and ability of the selling biotech to deliver on the deal.

Evaluating Return

In general, biotech/pharmaceutical deals have not been done based on hard return on investment calculations. The tendency for many biotechs to underprice the value of their technology or drug has made this difficult. It can also be difficult to determine precise value if the technology or product is not fully developed or so early in the drug discovery process that precise valuation requires too many assumptions.

The exception to this is the fully developed drug. A deal involving a fully developed and approved drug is easier to evaluate, since it relies on market estimates that can usually be done with some precision. These deals also fall into the clear area that pharmaceutical has interest in — more drugs to add to the sales bag. And usually the calculation of value will result in high valuations, sometimes multiples of a biotech company's market value. Multiple examples of these types of deals exist, such as the Pfizer-Eyetech deal. Many of these deals are or can become de-facto M&A transactions.

Risks

The risks associated with deals once signed are all about execution — the ability of the selling biotech to deliver on the terms of the deal over the timeframe, and receive the called out milestone and royalty payment earn-outs. The risks associated with technology working or a drug completing development are usually under the control of the biotech, but can require significant resources, especially if development takes longer than planned. And development usually takes longer than initially planned. These risks also drive the cost of completing the deal. It is usually important for businesses to make money on deals done. Selling the deal too cheaply may not cover the costs of unexpected delays, additional people, equipment, materials needed, additional testing, or any other resources required.

Some risks of execution are not under the control of the selling biotech, such as the buying company changing interest or business direction, or the buying company being acquired. It is necessary to deal with these risks in the contract of the deal. For example, there can be expensive drop out terms for the buyer if interest changes. Change of control clauses can be inserted to protect the rights of buyer and seller if either is involved in an M&A transaction. It is a valuable exercise to anticipate any and all contingencies, even if they are considered highly improbable, and try to insert clauses in the contract to deal with them up front. The clauses may not be acceptable to the buyer or seller in the deal, and may not make it into the contract, but they can at least be discussed and potentially negotiated.

Even if not covered, then the risk is at least understood and some planning can be done to ameliorate it if the situation changes.

So first, starting at the term sheet, ensure that the deal gives adequate value for the technology and at least covers the costs anticipated. For a small biotech company, even deals that appear to involve $10 to $20 million in payments might not be appropriate if the costs are $50 million to complete. This seems obvious, yet companies have signed deals like that.

Second, during the contract stage, insert sufficient safeguards to protect against all eventualities. For example, the customer pharmaceutical should not have an option to pull out at any time with no penalties. Ongoing issues can be addressed by a joint steering committee that meets on a regular basis. Establish at the beginning an internal champion at the customer, and place that person on the steering committee. It should be someone at a reasonable level of authority, known by the biotech and with significant incentives to execute the deal well. Good deal execution relies on ignoring the contract and setting up a good working relationship, until that doesn't work and you have to rely on the safeguards in the contract itself.

Golden Rules

First, make the deal a win-win situation, so both the biotechnology seller and the pharmaceutical customer are going to be happy with the results.

Second, establish teams on both the buying and selling sides that have a personal stake in the deal — people for whom significant personal achievement or failure is tied to the success of the deal.

Third, on the biotechnology side, assume the worst. Put a contract together that spells out both the risks taken in signing the contract, and the protections if the deal deteriorates. Ideally, of course, the goal is a deal so successful that nobody ever looks at the contract again. But it is still a good idea to make sure that if the deal ever falls apart for whatever reason, the biotech is protected.

There are numerous factors that can cause a deal to deteriorate. The companies might have a great relationship; then the buyer merges with another company. The priorities change, or the team changes. Suddenly, the situation is unworkable, having nothing to do with the supplying biotech. The reverse can happen as well. The biotechnology company can merge or be bought, and the deal can suddenly not make sense to the buyer. Usually, larger pharmaceutical companies are much more knowledgeable about their contracts, what is acceptable to them, and how they will protect themselves in deals. Smaller companies are much less aware and much less used to scrutinizing contracts. It also becomes difficult at the final stages of the negotiations when there is significant money involved — the pressure to get the deal done is huge. It is easy to downplay potential problems and just sign. This is the time when a flag should go up, and the consequences to the biotech if any such problematic events take place should be evaluated. The final rule: don't bet the company on the improbable but poor outcome of one deal.

Mr. Thomas Klopack was the chief executive officer of Aclara Biosciences, a public company in Mountain View California, from 2003 to 2004. In December of 2004, Aclara merged with ViroLogic, Inc. to create a significant force in the emerging field of molecular diagnostics. From 2002 to 2003, prior to joining Aclara, Mr. Klopack was the executive director of PhotoThera, Inc., a private medical device company in San Diego developing non-invasive treatments for stroke. From 1998 to 2002, he held the position of chief operations officer of Aurora Biosciences in San Diego. Aurora was a fast-growing and profitable biotech company focused on providing integrated solutions making drug discovery more efficient using innovative cell biologies, fluorescent chemistries, and screening systems. In 2002, Aurora merged with Vertex Pharmaceuticals in Boston. Prior to joining Aurora, Mr. Klopack held a wide variety of domestic and international management positions at Raychem Corporation in Silicon Valley, a high-tech company delivering enabling products to a broad range of industries based on cutting-edge material science.

Mr. Klopack holds a B.S. in chemical engineering from Carnegie-Mellon University and an M.B.A. from Harvard Business School.

Successful Biotech Deals

Gerard J. McGarrity, Ph.D.
President and Chief Executive Officer
Intronn Inc.

Business deals in biotechnology represent relationships that contain many of the ingredients of courtship and marriage. Sometimes until death do us part, sometimes ending in divorce with bitter custody battles. Courtships begin in a variety of ways, including blind dates arranged at partnering meetings, introductions made by family friends, as well as arrangements based on previous relationships between the families. Occasionally, the meeting between the parties can be completely serendipitous. Smaller companies have been known to be less selective — the struggling tradesman hoping to marry into the family of the town's wealthiest merchant.

Why do companies in biotechnology pursue partnerships? According to the Association for Corporate Growth/Booz-Allen & Hamilton, the three major reasons are to: accelerate growth, access critical capabilities, and enter new markets. Building critical mass, accelerating research and development, and cost and capacity reduction are also significant.

I have been with a small biotech company, a mid-sized biotech, and a large pharmaceutical company. While personalities can differ, there are clear characteristics that are common to all biotech discussions and negotiations. Many reasons can explain why a specific deal is not done—not the right product fit, inadequate ability to market, not meeting threshold revenues. The one ingredient each partner can control is the preparation. Complete and intensive preparation is the key to successful deals. The parties have to be completely immersed in the preparations, and in the deal. It cannot be a part-time enterprise. Adequate preparations will minimize, and hopefully eliminate, the surprises. In a perfect world, more preparation should reduce the number of surprises. For a deal to be successful, you need to know everything about it—inside and out. Remember, it can be a courtship leading to marriage. Know everything you can about your partner; it's where your company will be spending a significant part of its future.

Identify the common objectives. Establish common ground. This applies both to the business objectives, as does establishing a strong positive relationship with the party on the other side of the table. You need to focus on the business, not the personal issues.

Identifying key issues and ranking them by priority is also important. What items must you have, and what items are less important. Early in my career, I heard a story about some tradesmen from upstate New York who came to southern New Jersey to buy some farmland for future investment. The suburbs were starting to encroach on the more rural areas, so this looked like a good investment for the small group. Two representatives of the tradesmen came to the settlement without legal representation. There, for the first time, they put a list of twenty-four items on the table for negotiation. The sellers, their lawyer, and their real estate agent became irate for this last-minute surprise. To get the deal done, they aggressively went through the list item by item. The tradesmen got only four of the twenty-four items they requested, but signed the papers. The seller's lawyer came over to say goodbye to the tradesmen saying, "No hard feelings. I hope we weren't too hard on you." "Not at all," replied one of the tradesmen. "We wanted two key items on that list. You gave us four."

Recognize the deal-killers. Parties can be flexible in some areas, but not in all. When small biotechnology companies negotiate with large pharmaceutical companies, different forces are operating. What is the small company looking for? Validation for one. Also valuation. Frequently, the small biotech company is bringing new technology and new intellectual property to the table. When viewed in the perspective of the lifecycle of the product, the small company is proficient in the earliest stages, drug discovery and basic research. Larger companies have the additional expertise; pre-clinical development, regulatory submissions, clinical trials, approvals, sales, marketing, and distribution. The smaller partner will always be pushing to have the deal help them to be more proficient in the latter stages of the product lifecycle, such as booking sales. Don't let the deal die because you cannot get the opportunity to book sales. If the product shows early success and you have the ability to grow your company, you can always negotiate another day.

Deal-killers can be universal, applying to most, if not all, negotiations, and specific to the particular deal. Money and intellectual property are often at or near the top of the list. Money is the major drive. Who gets what? It's difficult to accurately calculate total dollars from what is publicly reported. The so-called BioWorld dollars assume all milestones are met. BioCentury has reported values for pre-clinical, Phase I, Phase II, and Phase III projects. Mean values reported were: $83 million, $268 million, $212million, and $277 million, respectively. Note the anomaly that a higher mean was reported for Phase I than for Phase II, which could have been ascribed to a single Phase I project that significantly skewed the values. These figures assume all milestones would be achieved, and include upfront/equity payments and milestones. The figures do not include royalties and/or sales and revenues. Development and sales milestones are included.

What defines a successful deal for you? What are your short- and long-term objectives? If you find that a deal can help meet or exceed your objectives, then you are probably looking at a good deal.

Litigation Issues

Intellectual property issues are critical components of most transactions, but especially those between a smaller company and a large pharmaceutical company. It is frequently the bag of intellectual property the small company brings that is a crucial element. One side of the table has to be sure the patent positions of the potential partner are sound, and that the combined entity will have freedom to operate. Is the technology the smaller company brings to the deal providing everything that is necessary to market the product if successful? The small company must also provide all intellectual property coverage necessary to make the union successful — but no more. Beware of the reach-through. One must protect against the partner gaining access to your technology that is outside the scope of the deal. We have turned down an otherwise attractive collaboration because of the potential reach-through to our background intellectual property. Our prospective partner said that was not their intention, but refused to insert language that would absolutely exclude that possibility.

One has to insure that the intellectual property needed to operate is clean, crisp, and clearly defined. As stated, the rule is to avoid surprises, and few things compare to the shock of learning that either you must access additional intellectual property, or that a competitor owns that intellectual property. That can add time and dollars to the anticipated time of revenue generation. I recall situations when what initially appeared to be rock-solid intellectual property was turned out to foreign applications but no United States patent application, or another situation when an impressive portfolio of licenses was for limited applications, not an omnibus license.

Also check the litigation landscape. Are any threats of litigation looming on the horizon, especially those that could threaten the viability of the company? The smaller company must be viable for obvious reasons. Will the smaller company be around for a long enough time to guarantee completion of the project? Does the company have sufficient funds to operate beyond the scope of this specific project?

The devil is in the details. Sometimes, what looks to be a very straightforward agreement turns out to be much more complicated when read word by word, line by line. One cannot over-scrutinize financial arrangements. In the end, that's why you're doing this; you are working for the shareholders, trying to increase valuation. The worst thing that can happen is going back to shareholders, after the consequences of a flawed deal, regardless of the size of your company, and saying, "Oops."

Ensuring that Biotech Deals Proceed Smoothly

Everyone involved in a biotech deal has the same objective: to try and increase the value of his or her company. Some people can be very aggressive, which can be both good and bad. There is a difference between being a skillful, tenacious negotiator, and argumentative. The primary goal is to achieve positive results for your company; being a tough negotiator is part of the process. The trouble begins when pride and ego get in the way. Check ego at the door. Nothing can dampen outlook quite like when the chief negotiators for the respective sides start off the discussion disrespecting the other side or the other negotiator. Or worse, when members of the same team are in significant disagreement. I've seen both happen. The latter should have been settled in rehearsals. If serious disagreements occur during the first date, there is little prospect of marriage. Poor interpersonal relationships turn the discussions into dour confrontations, chilling the atmosphere and affecting just about everyone in the room. I think particularly of discussions between a large pharmaceutical company and a leading academic scientist where the purpose was to renew a sponsored research agreement. The antagonism preceded the meeting, and neither the team leader for the company nor the principle investigator for the university would back down. The rest of the discussion was a sideshow, and the agreement was not renewed. Both sides lost. It works in the opposite direction as well. Good team members, well rehearsed and working together, can work wonders during a negotiation.

A good relationship can also work wonders. Harold Shaub, the former chief executive officer and chairman of Cambell's Soup related to me how he once achieved an acquisition. Negotiations had been going on for days, with lawyers, accountants, and marketing specialists for both sides at a stalemate. The chief executive officer of the smaller company was listening to his group of advisors, and the outcome looked doubtful. Frustrations mounted. Harold dismissed everyone from the room and sat down and asked his counterpart to list his concerns. It was a one-on-one encounter. They had a deal in less than an hour.

Typically, the other side will try to convince you that you are asking too much, and that you have overstepped your bounds. Go back to the basics. Preparation. If you are thoroughly prepared, you will usually have a good sense of were that line is located. Consider the possibilities if the other side is balking. The possibilities include that one or both of you miscalculated (the other side may not want to do the deal), or that they are motivated by something else. I recall a discussion where it became apparent that the other party was actually trying to manage my product in a negative way. They weren't interested in helping me move my product along; rather, they saw my product as competition, and they were trying to keep it stationary while they moved their own product ahead.

When you are working a deal, you want to see everything, and get full disclosure. The challenge lies in the fact that the definition of full disclosure can vary between the parties. In your planning, you must develop a full list of what you must review, and if it's not presented, ask for it. Special considerations apply to merger and acquisition transactions. Full disclosure is essential in regard to the science, finances, and intellectual property portfolio. Look beyond what they are presenting. In scientific reviews, ask the detailed questions. The results of the presented experiment look impressive. How many times was the study repeated? Were there failures? How reproducible are the results presented?

Patents must be scrutinized. You need to look at not only existing patents, but also patent applications, and whether anyone has ever issued a challenge to an existing patent. Patents are the key; they're an extremely valuable asset. Trade secrets are also very valuable. There are instances where a company's technology combined with its trade secrets can circumvent competition.

Marketplace Opportunities

Wood/MacKenzie reports that by 2010, in-licensed product sales will equal internal product growth contributions in the pharmaceutical industry. That is a significant change in the last five years. BioCentury reports that from February of 2003 through March of 2004, twenty-nine of fifty-two industry deals (56 percent) were discovery or early-stage deals. That is an even more significant change from the not too distant past, when large companies wanted only programs that were in Phase II clinical trials. The marketplace is always speaking; we just have to listen.

Intronn has benefited from this change in a small but, for us, significant way. In 2004, we began a new program to address the cardiovascular market. Specifically, we embarked on initial studies to insert, or trans-splice, the sequences encoding human ApoA1 into the highly abundant albumin pre-messenger RNA (pre-mRNA). The trans-spliced sequences would be under the control of the endogenous albumin promoter and regulatory machinery. Albumin accounts for 54 percent of plasma proteins, and is the most abundant transcript in human liver. Human liver produces approximately twelve grams of albumin per day. The objective is to increase the amounts of ApoA1, and therefore high-density lipoprotein in the plasma for patients with or at risk for atheroschlerosis.

When first we met with other companies at partnering shows and in other general discussions, we put the ApoA1 project clearly in its proper place — at the end of the presentation as an early-stage program with little laboratory data. Initially, it was a paper concept. We quickly discovered that many companies wanted to talk about this clearly early-stage program. The market was huge, and had been validated with the published studies by Esperion that the milano variant of ApoA1 showed effects in a Phase II trial. Pfizer added further validation by purchasing Esperion for $1.3 billion, although the price and the action were clearly linked to other cholesterol-lowering drugs in the Esperion pipeline. While Intronn has no deals to report, the discussion dynamics have been altered by the inclusion of the potential product addressing such a large market need.

The decision to pursue ApoA1 and HDL therapy was an easy choice for Intronn. However, at a general level, companies have to be able to spot an opportunity and be ready to respond to it. Is the new opportunity valuable enough to merit budgeting available resources to this new opportunity? In most situations, taking on a new project will not generate new resources. Resources will have to be taken from other ongoing or cancelled projects.

If the opportunity you're looking at is closely related to your core technology, the choice is easier than if it's not close to your field. There is a sense of comfort with the familiar. The larger challenge is to review a potential opportunity that is in a field outside your company's zone of expertise. How do you get up to speed quickly? It is possible, however, to get up to speed in a new industry or technology in a fairly short period of time, depending on the resources available. One can buy the reports offered by the business analysts in therapeutic areas, or in platform technologies. Alternately, good consultants are available to help you get to an acceptable level of knowledge. The reports companies submit to security exchanges, as well as the Web sites of companies engaged in the field, represent valuable sources of information. As with other areas, a network of colleagues can work wonders in educating the C-level executive and the company.

The bigger challenge is making the decision to enter a new area, one that is not in your usual comfort zone. Information gathering is a crucial part of the exercise. However, when the information gathering has been completed, it is still a decision based on facts, personal chemistries, and at least some instinct. No formula has been derived that can absolutely predict if a deal will be right for you or not. In the end, choosing the right deal will always have some component of art.

Dealing with Risks

A chief executive officer makes hundreds of decisions every year. A handful of these can affect the long-term viability of the company. Unfortunately, the chief executive officer doesn't always know what the long-term consequences of any one specific decision can be. This is generally not the case in business deals. The decision should be made on the merits of the case, although in some instances, the short-term financial rewards can exert strong influences on the other aspects of the deal.

Risk analysis is a critical factor in deal evaluation. Earlier, I stressed the importance of preparation. Prepare. Prepare. Prepare. However, even after complete and full preparation, there will be some element of risk in coming to the final decision, more in some deals than others. Harold Shaub told me another story. When Cambell's Soup wanted to launch a new product, they would do everything to get the market analysis — convene consumer and taster panels, complete market surveys, and introduce into small test markets. Yet after all these steps, the company could not always predict what the end results would be, even after all those preparations. Sounds like clinical trials, where a company can have impressive animal data, and encouraging Phase I data, and perhaps even Phase II data. At that point, the seller looks at the data and considers it to be a highly valued package. The buyer will respond, pointing out trials that looked good in early clinical development, only to falter in Phase III.

Conceptually, risk analysis is straightforward. In your preparations, your team defines where the relative threshold for acceptable risk is. In the reality of the discussions, that threshold gets blurred, and the equation changes during the discussions. The team must be able to update their definition of acceptable risk.

The type of risk is also relevant. Risk must be measured, both qualitatively and quantitatively. Beware of the case of low probability, high impact. You don't want to place your company on the table, risking survival of the firm, even for a situation that has low probability, unless of course, not to deal would have the same result occurring earlier.

Partnerships

The number of partnerships has increased dramatically in the past three to five years. Traditionally, smaller companies had to get their products into Phase II clinical trials before a partnership was signed. Not anymore. The same BioCentury article cited earlier showed that of fifty-two reported deals, twenty-nine were in the discovery or pre-clinical stage (56 percent); thirteen were in Phase I or Phase I/II (25 percent); and ten were Phase II and beyond (19 percent). The sea change is even more evident on the Web sites and presentations of large pharmaceutical companies. The Novartis Institute for Biomedical Research reports on their Web site that it had a total of 4,460 contacts in 2003. Of these, 3,800 were unsolicited, but 660, approximately 15 percent, were solicited by Novartis. A total of 296 proposals progressed to non-confidentiality presentations and led to 146 confidential presentations. Seventy-two term sheets were generated, netting twenty-two signed deals, or approximately 31 percent of term sheets were concluded. Overall, twenty-two of 4,460 proposals led to deals (0.5 percent).

Merck presents a similar, although less detailed, story on their Web site. In the years 2000 through 2003, the number of Merck's licensing deals totaled twenty-two, twenty-three, thirty-eight, and forty-seven, respectively. They reported that an estimated 3,000 external proposals were scrutinized in 2004. If the number of licensing deals stayed the same in 2004, approximately 1.6 percent of all proposals led to signed deals. There are now more partnerships between large pharmaceuticals and biotech companies working collaboratively in the early-stage development of products and programs. This is a fairly new phenomenon. In the past, the typical pharmaceutical company would want data from Phase I and even Phase II clinical trials conducted by a biotech company. That still occurs quite often, but not as much as in previous years. Today, there is much more activity going on during the preclinical phase.

The more intangible items focus on areas such as company culture, including size. Partnerships between large pharmaceuticals and smaller biotech companies are usually driven by the science — the leading partner is generally the one who has the most knowledge of the product or field. If both companies are on equal footing, then you need to look more toward which management style or company culture is better suited to the marketplace. Compatibility must be considered when you are looking at whether a large multinational pharmaceutical can work with a small, innovative biotech company.

A significant challenge occurs when you are entering a field where you have less knowledge than your prospective partner. You still need to know the territory. When you are looking to partner with a company, you have to know the territory and the marketplace. Do you have the in-depth knowledge necessary of the specialization that is involved in the product or technology? Define the expected synergies. It's important to assess whether those synergies will enable you to accomplish your goals, which in turn helps you to determine if the deal is right for you.

Mergers and Acquisitions

Mergers and acquisitions (M&A) are specific subset of biotechnology deals. I have been through three, all completely different and distinct. The first was the acquisition of Genetic Therapy, Inc. (GTI), where I was a member of the management team, by Sandoz (now Novartis) in 1995. Within nine months, Sandoz merged with CIBA to form Novartis. In that time period, GTI went from a 200-employee specialty biotech company to a being part of a 90,000-employee multinational pharmaceutical company. However, while we were affected by the merger to form Novartis, the GTI employees were mostly protected, since there was no similar type of activity on the CIBA side. The third merger occurred when I was with Cambridge Genetics Ltd. (CGL), a spinout from the Laboratory of Molecular Biology of Cambridge University. Cambridge Genetics acquired Cambridge Drug Discovery Ltd. (CDD) in 2000. The three cases had vast differences, primarily driving from the sizes of the partners.

GTI had two separate sponsored research agreements with Novartis, including a clinical trial in glioblastoma, where early results were encouraging. They knew us and respected the science. The initial discussions focused on a comprehensive review of the science. Every project was reviewed, from pilot experiments to clinical trials. They reviewed manufacturing, quality program, and patents. After they had comfort with the science, business discussions ensued. While different points were covered, the major topic was purchase price. After the basic science and associated fields were examined, along with the intellectual property, the board became involved. Purchase price was key to their return. GTI also hired an outside consultant who provided guidance during this phase. Several rounds of discussions followed, and several times we at GTI thought the discussions came to an impasse, since we did not hear anything for weeks at a time. Finally, in the first week of July, 1995, we came to an agreement. GTI was purchased for $295 million.

GTI provided entry into a field that Novartis had been interested in — gene therapy. GTI had the most experience in clinical trials and in manufacturing. The deal was good for GTI. It provided a profitable exit for the investors, and it gave Novartis a stronger presence in gene therapy than any other pharmaceutical company. Just months before the GTI purchase by Novartis, Chiron had acquired a gene therapy company in San Diego, Viaene. At GTI, we believed the relatively low purchase price that Chiron paid for Viagene drove our final purchase price down.

The situation at Cambridge Genetics was vastly different. Synergies were the key; both companies were small. CDD was a drug discovery company that performed high-throughput screening on contract. They had customers and a limited number of screens with additional screens in the pipeline. CGL had a strong intellectual property portfolio with applications in high-throughput screening and gene therapy. The willingness of GCL investors to put up additional cash to add a professional line of contract services as well as a state-of-the-art screening facility was a driving force. Chemistry was good, and the two management teams were committed to getting the deal done. The contracts were signed just months after the first discussions.

Negotiating Deals

As already stressed, people play a critical role in negotiating a deal. People can kill a deal. The first phase of the negotiations is typically science-driven. If the science is not there and if the science staff, especially the chief scientific officer, is not satisfied, closing the deal will be difficult. There has to be a champion. After the scientific review comes the marketing, financial, and legal, including intellectual property components.

The question of whether to use the chief executive officer, and if so, when, is a tactical decision. In M&A activities, the chief executive officer will have the final say. In most cases, the chief executive officer of smaller companies will be heavily involved in M&A activities, either at the table or in negotiations. Each transaction is different. However, in licensing deals, their involvement can be the final resort. On the other hand, if the chief executive officer is used early, the value of him or her as a final arbiter for the company can be lost. The business development team has to know when to bring in a chief executive officer, because if he or she says no to a deal, then there is minimal room for additional movements.

Making a Deal Work for You

Mistakes can occur, especially when adequate preparations have not been made. Letting egos get in the way can also cause serious problems. Runaway ego can often result in wrong choices and wrong decisions.

You cannot allow yourself be pressured into a deal. The other party may be in a hurry to finalize the deal. A mistake small companies can make when negotiating with large biotechnology and pharmaceutical companies is to attempt push the larger company to come to a decision. I have witnessed a small company tell their colleagues across the table that if they don't get a deal within a prescribed period, they will get one from a competitor. This represents a failure to understand the culture and the process of large companies. Think of the numbers listed earlier for Merck and Novartis. The process of dealing with several thousand proposals across perhaps a dozen portfolio areas is a tightly defined and controlled process. To tell, or demand, that the large company suspend its normal and standard procedure for this particular deal is at best counterproductive and at worst the kiss of death. Exceptions always occur, but if you sincerely have to play this card, be diplomatic and understanding. I have seen this demand result in an end to discussions.

Each deal and each negotiation is a learning experience. Every deal is different, and there is no standard operating procedure. Learn from your colleagues; learn from across the table. A database of experiences is one of the most valuable assets a team can have. In that regard, the business deal is different from marriage, where most of us get only one chance.

Mr. Gerard McGarrity, Ph.D. is president, chief executive officer, and co-chair of the board of Intronn, Inc. Intronn is his fourth early-stage biotechnology company. Prior to Intronn, Dr. McGarrity was founding chief scientific officer of Cambridge Genetics Ltd. (CGL). CGL purchased Cambridge Drug Discovery and was eventually sold to Biofocus PLC, now listed on the London Stock exchange. He was formerly was senior vice president of Genetic Therapy, Inc./Novartis. Novartis purchased Genetic Therapy in 1995. Dr. McGarrity remained at Novartis for three years following the purchase of GTI. He was chief executive officer of the Coriell Institute in New Jersey as well as adjunct professor of Thomas Jefferson University and Robert Wood Johnson Medical School. He was a member of the board of trustees of Thomas Jefferson University in Philadelphia and was given the school's Distinguished Alumnus Award. He served two terms on the recombinant DNA advisory committee of the National Institutes of Health, including one term as chairman.

During his academic career, Dr. McGarrity was a consultant to several biotechnology and pharmaceutical companies. He consulted for Genentech from 1984 through 1991. He led scientific delegations to China, Japan, and several European countries, and was a United States National Academy of Sciences exchange scholar to Czechoslovakia (now the Czech Republic). He is an honorary citizen of Bordeaux, France. He has been issued four United States patents and has authored or co-authored more than 160 publications.

Biotech Deal Strategies

Mark Schwartz, Ph.D.
President and Chief Executive Officer
Bayhill Therapeutics

Introduction

My strategy for doing successful deals in biotech is to ensure that the deal is aligned with the strategic components and strategic direction of the company. One component would be to eliminate all the extraneous issues of why you do a deal — because the chief executive officer wants a bigger company, or because the board might be pushing you into certain deals. You do a deal because it's critically aligned with the most important strategic components of the company. Doing a deal of any kind — a merger or acquisition (M&A), for example — requires considerable resources in a company, especially a small company. If it is a small company, it can tie up management, your key scientific leadership, and people at the bench preparing data, doing experiments, or cleaning the lab for walkthroughs. You want to make sure the deal is critical to where you are going, and not just part of building territory or some perceived critical mass that really is not going to move the company forward. This is especially true for a small company, where there is only a limited number of times you can go out and do an M&A or an end-licensing deal.

Measuring Success of a Deal

A deal's success is relative to what it is you're doing. If you're trying to do a partnership around a therapeutic, you need to outline the key strategic elements. Are you looking for the vanguard strategic relationship to say we've done a deal with a Merck or a Lilly, for example? Are you looking for financial support, development expertise, or marketing support?

If it's an M&A transaction, are you looking to bring on specific technology you don't believe you can build yourself? Are you looking to bring in a late-stage clinical compound, for example, because you're an early-stage discovery company?

The success of a deal is dependent upon the first question: What is the strategic focus of the company? You should be able to whittle it down to just one or two very key things. For example, do we want to be a late-stage clinical company? If so, I'm going to look for a deal with a company that brings a late-stage clinical product in an area we know something about; an area in which we can move forward with the skills, expertise, and financial resources we have here now, and that is going to make us truly a late-stage clinical company.

If you're looking to partner a drug, what are you looking to get out of it? Is validation by a partnership with a marquee pharmaceutical company a priority, or is maintaining development control the priority? Do you want to maintain involvement in the development of the compound, or are you merely looking for a lot of cash? The success of the deal should be measured by these strategic priorities.

Negotiating

There are two general phases to most negotiations. The first is the management team sitting down and agreeing on: "What do we want to accomplish. What are our strategic goals? Who do we believe has what we want? With whom can we construct a win-win situation?"

If it's a therapeutic deal, does it have to be a big pharmaceutical company? If so, determine which of them are playing in this area. Then determine who has the strategic focus or knowledge of the explicit technology you're working in, and approach them.

One also has to be open to opportunistic approaches. In the case of a small, private company like ours, board members, Scientific Advisory Board members, and investment bankers will all approach us with different opportunities. One needs to have a very high bar of what one engages in, or else you can end up chasing a plethora of ideas that don't completely map onto what your strategic goals are.

We have set up a process where our business development people will keep in contact with everybody in the field, because there's a lot going on out there in the world, and there are a lot of interesting ideas and opportunities that may not be otherwise apparent. However, we are very careful about absorbing corporate resources and continued in-depth discussions on opportunities that do not clear the initial hurdle. We listen as much as we can without getting too wrapped up in going to the next step unless it clears the key hurdles we need to make a deal happen. Does the transaction forward out key strategic goals? Is there synergy in the outcome? Is it a win-win for all concerned?

On the partnering side, just like with an M&A, the company needs to have clear objectives for the transaction. If you fail to establish those objectives beforehand, you will end up working on transactions and going down the path the wrong way until it falls out for any number of reasons: your partner may not be willing to pay enough, they're not going to give you co-development rights, or they may want to have a joint steering committee, etc. In this case, a significant amount of time and energy has been wasted, and a significant opportunity cost, in terms of pursuing the right opportunity, has been lost.

Roles and Motivations of the People Involved in Negotiations

The second phase of the negotiation is determining the roles and motivations of the people and organizations involved. In order to determine the ultimate goals and objectives of the party you are working with, you must listen carefully to what's put on the table and pressure-test everything you hear.

From our side, we have a team of people involved. I'm the chief executive officer, but I try not to be the only negotiator. In fact, I have tried many times not to be the lead negotiator, because other people's perspectives bring great breadth to our internal analysis and decision-making process. I prefer to be called in when there's a critical issue; having to deal with their team, their chief executive officer, or their lead negotiator who has been brought in already on that critical issue allows one to pressure-test the assumptions one is making about why they're doing this deal.

One of the reasons negotiations take a long time is because it takes time to really understand what the other party wants and what they say they want. It's not that anybody is ever misleading, but are you interpreting it the same way? That is particularly true in an M&A, because there's rarely a merger. It's called "mergers and acquisitions," but the vast majority of time, it's really an acquisition; there's usually a dominant party. Understanding what the dominant party really wants to accomplish takes time, and is ultimately key to a successful transaction.

Spotting Opportunities

There are several ways to spot new opportunities: press reports and press releases, scientific and technical meetings, business and professional meetings. We get some leads through investment bankers and many contacts through outside investors. We are always open to talking with other companies and scientists on an introductory level, because you never know where the opportunity is going to come from. We are, however, very discriminating as to which opportunities we let come into the company and start absorbing people's time.

On the partnership side, identifying major collaborators is more straightforward. Most companies we would look to partner with are already working in our field, and it is unlikely that a major pharmaceutical company is going to jump into a brand new therapeutic area. So, we can start our prioritization by our own knowledge of the industry.

In general, our transactional strategy is a broad-range approach of actively doing your homework on companies in the field and keeping abreast of key people and scientific events. Also, I personally and actively solicit from our board members, our scientific board members, our consultants, and our founders any opportunities they think are relevant. As a small company, we get a lot of M&A opportunities through them, in particular through our investors. There are only about fifteen or twenty big pharmaceutical companies, and it's not very hard to identify the appropriate ones. Venture investors tend to have a deep exposure to smaller companies and opportunities, and they are a rich source for contacts there. If you're open and you actively solicit contacts, you can get a constant flow of opportunities coming in to your investors.

How to Get Information about a Potential Deal

There are a number of different ways to get information about specific deals. Let me talk specifically about Bayhill Therapeutics. We're a company with a deep expertise in autoimmune disease, immunology, and inflammation. We know those disciplines, because that's what we do.

Our scientists are constantly going to meetings and reading papers, so they naturally keep up with all that's going on. I, our business development head, and other people are constantly reading the business press in this area, so we know who has drugs in the clinic and who is working in this area. There's not an explicit set of tools we need to utilize to keep up with this area.

If suddenly we wanted to get into a new field, say, cardiovascular, what would we do? We would probably use some of the common tools, the Internet being the initial screening tool. You can also buy research reports, although in general, my experience has not been overly positive with many of the research reports out there. If you're new to an area, it's probably not a bad way to go. If you're steeped in the area, I don't find it particularly useful. We would rely very heavily on consultants and our scientific contacts. Bayhill has a very active group of founders from Stanford University that are extremely successful academic scientists. Their contacts, collaborations, and connections in the academic research area run very deep in a variety of areas.

Our ability to talk to some of the leading academic researchers in the field is very open to us. We use that often to get the leading researchers in the field to come in and tell us about a specific area. We rely on them a lot, because we're very science-driven. We start there, and that's probably the one area that others may not use quite as extensively, because we do have those contacts in the field.

How to Evaluate Potential Return on Investment in a Deal

For a company like Bayhill, we evaluate a deal's potential return on investment by basic financial modeling revenue, net present value, and return on investment in a very traditional manner. However, in an early-stage biotech company, from my perspective, there's a more relevant binary of analysis: probability of clinical success. We're developing a drug; it takes eight to twelve years and $500 to $900 million to develop one drug. The return on investment from a strict financial analysis is far less meaningful than what your shot of getting this drug to be successful through the regulatory process and ultimately to market is. Who's going to be the best partner to make you do it? Where are we going to get the money to make it happen?

The markets we go after are very big. Whether it grows by 5, 10, or 20 percent, it's not going to make a huge difference in the analysis, as the scientific and medical risk is a much bigger factor for us than the revenue risk. If we are successful clinically with our program, and we get FDA approval to market our drug, we will have a very successful product, and a successful company. It's a binary situation for us: the drug will be approved or not be approved. Therefore, we have to try to optimize any partnership or M&A from a perspective as to what will further the development success of the drug.

If we're in-licensing in a drug, or buying a company because they have a program that's in late-stage clinical trials, it's much more a scientific analysis than a financial analysis. Not that we don't do financial analysis, but unlike a technology company buying an existing product with a revenue stream, or a manufacturing company buying a manufacturing plant, the scientific merits and clinical risks are predominant in our analysis.

Dealing with Risk

In the therapeutics arena, and in the biotechnology area in general, for an early-stage company like ours, the risks are almost all scientific and clinical. The question is: Will the drug make it through the regulatory hurdles or not? Will this technology work? The risks are really dependent on the scientific data that support the technology, the product, or the company.

We do extensive scientific due diligence on any deal we are considering. I don't think there's any magic to it. You do as much exhaustive scientific analysis as you can, and you try to find the absolute best people in the field to help you analyze those risks.

If we're looking at safety and toxicity, for example, we'll try to find the best people in the field. There is no substitute for world-class expertise. We can model potential success financially, but that is a much less important factor than making the right scientific decision. If you do a deal with a company in a new therapeutic area and the drug fails, you've lost it all. If the drug is successful, you've made it to market, and the market's probably going to be reasonably big; if that's what you thought it was today, it's probably going to be big in the future as well. Risk for us is really a scientific question.

Best Advice for Doing Deals

The best advice I've ever had about doing deals is to focus. Deals can absorb a lot of company time, money, bandwidth, energy, and psychology. I've been involved in doing deals we have spent a lot of time getting wrapped up in, when in the end they fell apart because they did not really add value to the company's strategy.

If you've made the decision that it's a particularly meaningful deal, it winds up with a strategic perspective, you've thought through all those components, and you believe it's truly going to add value to the company moving forward, then go do the deal. But most deals that get started don't get done. The reality is that probably less then 5 percent of all initial discussions come to fruition.

Prior to joining Bayhill Therapeutics, Mr. Mark Schwartz, Ph.D. was chief executive officer and president of Calyx Therapeutics, where he led a significant fundraising round and the development of novel anti-inflammatory compounds. For twenty years, he has held research, product development, and senior management positions in the bioscience industry. Previously, Dr. Schwartz was with Trega BioSciences, where as the chief commercial officer, he played a leading role in the development of the company's iDiscovery strategy and in the merger of Trega with Lion Bioscience AG. Dr. Schwartz held senior management roles at Synteni/Incyte Genomics, Inc., where he played a key role in the successful merger of Synteni and Incyte Genomics, and at Tripos, Inc., where he pioneered the accelerated discovery services business unit and was responsible for the development of the company's combinatorial and drug discovery business. Earlier career positions included product management and product development positions in diagnostics at DuPont, and protein instrumentation at Applied Biosytems.

Dr. Schwartz holds a Ph.D. in biochemistry from Arizona State University, and he received his B.A. in chemistry from Grinnell College.

Secrets to Success

Abhi-Shek Jain
Chief Executive Officer
WTP Capital, LLC

ASPATORE BOOKS
C-LEVEL BUSINESS INTELLIGENCE

The Secret of My Own Success

My definition of a successful deal is one that, after investment, meets all expectations, including expectations of operational efficiency, market traction, growth, profitability, value creation, marketability, and exit. When expectations are met, everyone is happy. Investors' wealth is enhanced, the fund manager's credibility and goodwill in the investor and financial community is bolstered, the investee firm's employees are assured sustainable jobs, and the market's needs are well served.

There is no proven formula, however, for ensuring success in a deal. Even the best thought-out investments can turn sour overnight. Having said this, my strategy for doing deals is three-tiered: research, evaluate, and structure (after deciding to do the deal).

I find it important to know enough about the deal before proposing to put investors' hard-earned money into it. Most successful endeavors are the result of 80 percent planning and 20 percent execution. Exactly how compelling an investment proposition is depends on what I know about how it operates, what its core competencies are, how unique its product or service offerings are, where it stands vis-à-vis competition, who its clients are, how vast its potential for growth is and in which areas, who runs the business, who's behind it, who does business with it, what incentives or restrictions apply from any government entity or regulatory agency or banking institution, and so on.

Until I am comfortable that what I know about the deal is enough for me to make a solid conclusion, I will not stop mining information. Information sources could include the Internet, broadsheets and publications, books, industry reports, analyst forecasts, interviews with deal proponents, office and plant visits, brief discussions with rank and file, talks with experts on the technology or on the industry, discussions with colleagues and other venture capital firms, and so on.

In the evaluation phase, my main considerations for deciding to pursue a deal are:

- the track record and management philosophies and practices of the firm's proponents
- the uniqueness of its product or service offerings and their relevance to the firm's target markets
- the company's potential for growth within its identified market niche(s)
- the roster of existing and potential clients
- opportunities for enhancing operational efficiency
- exit options
- the likelihood of generating meaningful returns throughout the life of the investment

If the deal rises to the surface and is deemed worth investing in, it now becomes important to structure the investment so that if it happens, any loss incurred is not so grave. Protecting the investment from bearing the brunt of any adverse eventuality — which, in turn, may cause it to erode in value — is just as critical and fundamental to the success of a deal as growing it to vast proportions. It is also part of our fiduciary duty to our investors to manage their wealth with prudence and skill.

One must bear in mind that though preparing the stage for a successful deal entails a lot of hard work, achieving it is not always the natural consequence. What is important is that possible risks are foreseen and provisioned for as much as the anticipated returns.

Making Decisions

In my experience, spotting opportunities for investing requires a keen understanding of what's going on in what marketplace, and an analytical ability to process the goings-on in terms of how they impact the present and the future. If I determine that a particular industry is attractive, I begin to do the research and to survey the field for possible investment opportunities through my network of friends and colleagues. Referrals are the primary source of deal flow; "cold submissions" are rarely that good. There are quite a number of opportunities to be tapped. But deciding which ones to exploit depends to a large extent, I believe, on what one has to bring to the table. I do not act on ideas I cannot bring value to. I do the type of deals I do because they are within my sphere of expertise or exposure. And the ideas come as I go about my day-to-day activities while keeping my ears to the ground. This is why I tend to narrow down to deals whose businesses I understand. If they are not within the food chains or the industries and markets I am familiar with, they do not form part of my considerations.

Granting the deals pass this first criterion, I tend to further discriminate towards deals in a nascent field or niche, as these have better prospects for growth than most. The specific companies in such a niche that appeal to me are usually those that have potential to scale and achieve operational and cost efficiency, that are managed by a credible leadership with a track record of success, and that may generate meaningful returns on my investment.

To keep up on a specific sector or company, we follow industry publications, attend seminars and venture fairs, and review proprietary research reports. But the best value seems to come from keeping your ear to the ground and listening to what industry executives, research scientists, and other investors are talking about at networking events and other gatherings. The informal word of mouth can often be illuminating.

The Art of the Deal

The most effective way to determine the roles and motivations of people I am transacting a deal with is to read and ask. Read their executive summaries or business plans in view of their past and current operating performance. Understand the capital structure and proposed use of required funding. Ask as many questions as needed—not all in one shot, but in bits and pieces, under more than a few circumstances—until you are able to paint a picture of what really drives them. More often than not, what I read and what I hear are sufficient for me to determine the motivations that may lie behind a deal. But really, what matters more to making an investment decision, for me, is not so much knowing the true motivations of the proponents, but oneself being convinced, to a comfortable degree, that goals and expectations for doing the deal will be met regardless. Whether a proponent wants to be bought out because they're burnt out from running the business, or simply wants to make a killing from an attractive valuation is of little significance in view of the future business prospects of the firm.

In my experience, the best way to approach negotiations is to come prepared with the facts, and with a clear perspective at the outset of what you want and what you are willing to give up, given your strengths and limitations, as well as those of the potential investee company. Since negotiations tend to require some on-the-spot decision-making, I find it important to listen well and be attentive to the other party's demands, to process the new information intelligently, to give in to reasonable wants whenever possible, and then to leverage these concessions to gain ground with your own terms.

A win-win situation is ideal. It is good to bear in mind, though, that the key to successful negotiations in the long run is the willingness of both parties to give and take. This means building and keeping a relationship of mutual respect. Part of this means being straightforward and not trying to "game" negotiations by springing more and more draconian provisions in the documents at the eleventh hour, knowing that the cost of walking from the deal gets higher for the investee company the longer they have been working on the deal. I feel all negotiations should start with a basic term sheet that sets out the key terms of a deal so there is a basic meeting of the minds early on. The terms should always be within market parameters; overly aggressive tactics generally don't seem to bring long-term value.

Return on Investment is King

At the end of the day, my function as a venture capitalist is to deliver superior financial returns to my investors. Success in a deal ultimately boils down to the numbers — what kind of return on investment was achieved and what the internal rate of return was.

I evaluate my potential return on investment by conducting standard financial analyses on the proponent's, for example, five-year financial forecasts; taking into consideration the company's assumptions on the prospects of the business, its historical financial performance, the likely exit, industry and peer price multiples, and my required hurdle rates, among other things. I find it essential to sensitize management's projections and subject the assumptions to certain stress tests (e.g., reduce sales growth rate, increase costs, assume delays in production or delivery, assume longer sales cycles, etc.) in order to cushion the effects of possible future adverse developments in the business or industry and to arrive at more conservative estimates of potential returns. If, despite the sensitization, returns remain exceptional, then the deal is more attractive. If not, I may pass on it. For me, six times would be an acceptable return for an investment held over a period of five years.

Dealing with Risk

I recognize that risks can never be completely eliminated in business as in life, in general. Often reminded of this fact, I deal with risks in a pragmatic way: I look for the risks as ardently as I seek the opportunities, and determine how each and every one of those risks may be mitigated. The risks worth taking, for me, are obvious: the common and the unlikely. Common risks are those that face most companies in a given industry (e.g., the risk of obsolescence in the high-tech gadgetry field, or the risk of being denied FDA approval in the health food industry, or the risk of theft/robbery in a twenty-four-hour convenience store). Risks I consider unlikely to happen are those with a statistically low probability of happening, such as those resulting from fortuitous or unlikely events (e.g., the risk of a building collapsing from an earthquake or a terrorist attack, or the risk of a luxury liner sinking in the middle of the Pacific Ocean, or the risk of mass resignations in a semiconductor assembly plant). These types of risks are worth taking.

Those that need serious consideration, however, are the risks that usually affect the likelihood of a successful or positive outcome if incurred. Examples are the risk of failing to meet targets and milestones due to incompetent management, or of losing clients through poor servicing or the lack of innovation in product line, or of loss of a unique value proposition with infringements on designs and concepts. These risks are not worth taking if the likelihood of their occurrence is high and little can be done to mitigate them.

When, at the outset, one is clear about the criteria and the parameters within which a deal should be pursued, it shouldn't take long to decide its fate. If, for example, it is not within my mandate to invest in armaments, cigarettes, liquor, or casinos, then any deal in this area goes straight out of the door. If the deal is within my sweet spot—say, in the biotech industry—I run through a quick list of basic features the deal must have and assess whether I should take it a step further. As mentioned earlier, the usual considerations for evaluating a deal are its line of business, the uniqueness of its product or service offering, its addressable market and client mix, its management profile, its value proposition and comparative advantage over competition, barriers to entry, its growth prospects and exit potential, and its expected returns.

I usually start with an analysis of management—if I am not comfortable that there is a management team with a solid track record in business (not just biotechnology), I look no further. In real estate, they say the three most important things are location, location, and location. In investing, I think it's management, management, and management.

If I get beyond management, then I look at the financials. If management's forecasts result in potential returns below my required level, the deal doesn't stand a chance, and I would care little about considering the other criteria. However, if management's forecasts produce high returns, but upon sensitization result in reduced returns at minimum or below-minimum acceptable levels, the deal will not survive either. Now if returns prove significant even after sensitization or stress testing, then the other criteria begin to play a role in turning the table for or against the deal. If, for example, after some research, a few visits, and some interviews, I am not convinced the technology is solid enough to steer the company in the direction its numbers point, I could walk away from the deal. If the fundamentals are good, but visibility for a smooth exit is dim, I could also walk away from the deal. If everything works, but for some reason the personalities of the management team conflict with my own ways of doing business, I would still walk away from the deal. As you can see, it is easier for a deal to be dropped than for it to make it to our banks.

Making Mistakes

Among the most frequent mistakes made when doing a deal are:

- Rushing to invest in it without first adequately evaluating its viability because you're under pressure to invest the entire fund, the investment is in a hot and sought-after space that everybody else has gotten into except yourself (e.g., the dotcom craze), or an investor or fund advisor is keen on pursuing it.
- Paying too high for a deal because you didn't have sufficient ammunition (information) to bargain it down to its fair price (e.g., investing in high-technology startups in China in the 1990s with (1) little but a few notions about the possibility of trailblazing, for example, integrated chip design for communication devices, or software development, for a potentially huge China market — its appetite for such products not fully understood yet assumed significant down the road, given the overall global trend at the time; (2) only western high-tech ventures against which to benchmark valuation, and (3) twenty-twenty vision hindsight.)
- Buying into the hype of "disruptive technologies." Few technologies actually are disruptive, so really understanding how unique a technology is can be important.
- Misunderstanding the difference between technical viability in the laboratory and technical viability in the market. Many technologies that work great in the lab are difficult to commercialize effectively.

How to Not Make Mistakes

Practice makes perfect. This is the sure and tested way of coming closer to investing in successful deals than not. A good track record as an investor boils down to the annual internal rate of return on investments measured in liquidity.

One should realize that low aim, but not failure, is a crime. Investors should put money into what they know well. Don't try and jump onto the latest bandwagon unless you really understand the business.

The Five Golden Rules

These are my basic rules for doing deals:

- Invest in what you know.
- Do your homework/research.
- Challenge all assumptions.
- Plan for the best and the worst. (Companies often fail because they don't protect against worst-case scenarios or because they can't scale fast enough to meet customer demands effectively.)
- Be fair in your approach. (Don't get greedy, and be straightforward in your negotiations. You have to live with the people on the other side of the table after the deal closes, so make sure they are satisfied, too.)

Mr. Abhi-Shek Jain is the managing director and chief executive officer of WTP Capital, LLC, and is responsible for the investment strategies and day-to-day operations of its funds. He has lead investments in companies such as Cibernet, Part 2 Technologies, B2Tecs, Vigilar, and Coderyte, among others. Prior to co-founding WTP Capital and becoming its chief executive officer, Mr. Jain served as the president of Megasoft, a publicly traded software services company that employed a cross-border service model. Previously, he was a founding partner in the law firm of Jones Jain, LLP, which was acquired by Greenberg Traurig, LLP in 2000. His practice focused primarily on the technology/business services sector, from representing startup companies on organizational issues and venture capital to advising major corporations on their international market penetration strategies. His clients included AOL, Intel, Painewebber, Morgan Stanley, JP Morgan, Emerging Technology Solutions, and many others. Mr. Jain currently sits on the boards of several private companies and is a governor-appointed member of the Virginia Biotechnology Research Park Authority and the lieutenant governor's business advisory board.

Mr. Jain also served as counsel to the Embassy of India and as the executive director of the India Centre (an Indian cultural organization promoted by the government of India and chaired by the Indian ambassador to the United States). He also has been the chairman of the American College of Technology Lawyers. Additionally, he recently served as the co-chair of the international team of consultants retained by the government of India to draft India's legislation on electronic commerce. Mr. Jain is a regular speaker on venture capital issues and is also the president of the ICEO Council, the largest chief executive officer organization on the east coast of the United States. He is a charter member of The Indus Entrepreneurs, an international organization of Indian executives. Mr. Jain has served as an adjunct professor of law at the University of Maryland Law School and as a lecturer in the University of Maryland's Smith School of Business. He has also served as a member of the editorial advisory board of Euromoney Law's Privatisation Opportunities *magazine. He also has the distinction of being the only venture capitalist selected by the United States Agency for International Development to advise it on market development and capital growth in developing markets.*

Finding and Capitalizing on Opportunities

Alexis V. Lukianov
Chairman of the Board and Chief Executive Officer
NuVasive, Inc.

C-LEVEL BUSINESS INTELLIGENCE · ASPATORE BOOKS

The Key to a Successful Deal

Deals in the biotechnology area take a wide array of forms, ranging from fairly straightforward licensing or cross-licensing of technology to outright acquisition. First and foremost, it is essential for a biotech company to seek products or technologies that will move the company forward and give it either a strategic or financial edge in the market. Enhancing a product portfolio or entering a new market are worthy goals, but leapfrogging your competitors' technology and market position should be a driver in most deals. Our company looks to do deals that complement our existing business and culture, and have the potential to significantly advance our market position. We weigh these goals against the time and resources that would be required to organically develop the same product or technology.

Our company is a medical device company focused on the design, development, and marketing of minimally invasive products for the surgical treatment of spine disorders. We strive to be the market leader in surgical access technology, as well as to provide all of the implants needed to complete the procedure. We take a very careful look at all of the different market segments and sub-segments within our rapidly growing orthopedic industry to find various products that could improve or expand our product portfolio. We also seek to enter new or unique market segments by developing or acquiring new technologies and evaluating numerous opportunities such as software-driven electronics, motion preserving devices, and service arrangements. As a business dedicated to the rapidly evolving spine market, we feel compelled to take a broad view of ways to accelerate our growth. We even track many companies outside of our present product scope to ensure that we are thinking about solutions outside the box and not relying solely on our existing product lines and related approaches.

What Sets Biotechnology Deals Apart

The two main areas within biotechnology are the medical device area and the pharmaceutical, or drug, area. One of the defining differences between the two areas is that the pharmaceutical side is more willing to make more, and sometimes even multiple, alliances with companies in the same sector, which is in contrast to typical arrangements on the medical device side. Companies on the pharmaceutical side also tend to position themselves early on for sale to a larger company, and they often look for strategic investors to help develop the business and potential markets. This can be driven by lots of factors, such as difficulty in gaining clear intellectual property protection, the need to rely on a larger partner for distribution of the product, and the expense and difficulty of developing a sales force.

In both areas, there are often companies that position themselves to be sold early in the company's lifecycle. This is often done to avoid the protracted and costly process of building the business. However, medical device companies seem to more often attempt to independently build fully operating businesses, complete with product supply, distribution, and marketing. Establishing corporate alliances and joint ventures for purposes of product development and commercialization is more common on the pharmaceutical side.

Finding Opportunity in the Marketplace

It is extremely important for any biotechnology company to follow its market dynamics and be aware of changes and emerging trends. One helpful method of learning about future changes, especially in the medical device area, is to consult directly with surgeons to get the information on changing needs or new opportunities. If there is a new device doing well in the market, or pre-clinical activities or clinical trials are generating significant interest, then we try to determine what problems surgeons are trying to solve in terms of symptoms and surgical procedures. Most of this information can be gathered from medical publications, presentations at scientific meetings, discussions with surgeons, and by keeping a watchful eye on budding companies. In the spine industry, clinicians are an excellent source of information, because virtually every new product under development was either developed directly from significant input from spine surgeons, or at least used by them clinically. We find there are small networks of surgeons contributing to deals and development efforts on an ongoing basis.

Evaluating Potential Return

A number of different perspectives help a company to evaluate potential return on an investment. Major areas of discussion and evaluation include the size of the market that will be penetrated, anticipated revenue ramp over the next five years, attractive gross margins that will be obtained, and the expenses associated with bringing the targeted product to market, including those associated with intellectual property protection and legal and financial due diligence. Regardless of the product's development stage, most deals require an injection of capital that needs to be accounted for. Attention should also be paid to the potential of the acquired business to continually improve operating margins through cost efficiencies or otherwise. It is also necessary to create and run various financial models to estimate the net present value of the deal and understand what your cash flow would look like while supporting a product at different development stages. From there, it is important to sort out what else will drive the costs of the deal. Once everything in the deal makes sense, it is time to move forward.

Risks and Red Flags

Dealing with risk comes down to instinct and experience within the marketplace. In addition, diligence also plays a major role in cutting risk. You won't hit every single ball out of the park. But it is possible to protect your company from many risks by being diligent and by thoroughly understanding the parameters of any potential deal. Each biotechnology company must determine its level of aversion to risk. A company like ours that is looking to increase revenue at a very high annual rate will probably be a little less risk averse as it considers deals that have the potential to greatly improve the company's market position. Risk aversion comes down to the profile of the company, the market in which it operates, and the management team.

The individuals directly involved have a significant impact upon the success of a deal. Even while completing their other due diligence, companies in the process of doing a deal must ascertain whether or not they are dealing with honest people who would be truly committed to the success of the combined company. When you start to see inconsistencies in what the other company claims to be selling, such as changing sales forecasts or uncertain intellectual property protection, this should raise a red flag. It is helpful to pay attention to how the deal was originally pitched and then to observe what happens on the second, third, and fourth round of discussions. As the discussions unfold, if the company starts to back off from some of its original claims, it is usually indicative of a pattern that will continue to complicate the deal. If a red flag indicates that a company is overselling their products, then it is time to look at everything from a much more critical perspective.

Mistakes

Courting a deal for too long leads to problems. It is generally ideal to complete a deal fairly quickly. Whether or not it is possible to efficiently close a deal comes down to financial analysis and sorting out all of the key deal terms. It is helpful to determine, from the beginning, if both companies take similar, productive approaches to solving deal-related problems. It is also important to honestly and critically analyze how far apart the companies are on all material items of negotiation. By understanding this information right away, it is easier to proceed smoothly through the deal. Many people do not take this step until late in the negotiations, because they are afraid of losing the deal. At this late point in the negotiations, companies too often find that they are so far apart in their thinking that the likelihood of completing the deal is low. To avoid this situation, our approach is to negotiate a rather detailed term sheet with a clear, written understanding of the major deal points before moving toward providing documents toward a definitive agreement.

Financial Strategies

Saving Money on Deals

The possibility of saving money on a deal depends on the area of focus and the financial structure of the deal. Certain deals, such as technology licenses, may be structured with nominal upfront payments and royalties to be paid only as sales are generated. A full asset or intellectual property acquisition may require significant upfront payments, with all later sales revenue accruing to the purchaser. The ability to save money is a question of realizing the potential synergies and generating incremental revenue. Depending on opportunity cost, it may be beneficial to spend more money up front to gain complete ownership and control of the desired technologies. It is possible to save money in the long run by acquiring technology that allows you to quickly enter the marketplace, rather than technology that requires significant additional research and development expenditures. By getting to the revenue stream faster, and potentially taking advantage of current market dynamics, there may be meaningful long-term savings.

Additionally, finding and taking advantage of the synergies and the points of consolidation are key to saving money on a deal. Synergies depend upon a number of different factors. They typically play a role in an all-out acquisition. In the case of buying an operational entity, an obvious money-saving tactic would be to consolidate on the distribution front. Melding some of the operational functions together, depending upon whether they are independent or direct sales organizations, is a first step. Then it is wise to look at the geographical differences or the physical locations. It might make sense to either sustain both locations with a skeleton crew, or to absorb the acquired location into an existing facility. In most cases, it is more efficient to combine the locations unless there is a strategic reason for keeping the second geographical location. Combining locations cuts out the fairly significant overhead expenses of a second facility.

In another strategic move, it usually makes sense to absorb the research and development portions of the acquired organization, thereby doing more development work for less money. From a finance and accounting perspective, it is cost-efficient to put everything in your own headquarters or subsidiaries if possible. Reporting from a remote location can be costly, so avoiding this is a money-saver.

Similarly, synergies can be realized in sales and marketing efforts, especially if the acquired products or technology complement existing lines. Many sales and marketing expenses are fixed, so the addition of new products or lines can be achieved at a marginal additional cost.

Mr. Alexis V. Lukianov has served as president, chief executive officer, and a director of NuVasive, Inc. since July of 1999. Mr. Lukianov was elected chairman of the board in March of 2004. He has led NuVasive from the startup phase as a venture-backed private company through product commercialization and to its initial public offering in May of 2004.

Mr. Lukianov has over twenty years of successful experience in the orthopedic industry, with over fifteen years in senior management. He held several top-level positions at Medtronic Sofamor Danek, the world's leading spine device company, including serving as president of Danek Medical and as president of Sofamor Danek USA, where he was a major contributor to the company's rapid growth. He was a founder, chairman, and chief executive officer of BackCare Group, a spine physician practice management company, which he sold to a Robert Bass company. Mr. Lukianov served as a director of the Ilizarov business unit of Smith & Nephew Orthopaedics, bringing revolutionary limb lengthening technology out of Russia and establishing worldwide manufacturing, marketing, and sales. He was also a key manager in a sports medicine joint venture in artificial ligaments between Stryker Corporation and Meadox Medicals.

Mr. Lukianov attended Rutgers University and served in the United States Navy.

ASPATORE BOOKS

C-LEVEL BUSINESS INTELLIGENCE